THE INVISIBLE FIRE

THE INVISIBLE FIRE

The Story of Mankind's Victory Over the Ancient Scourge of Smallpox

By
Joel N. Shurkin

AN AUTHORS GUILD BACKINPRINT.COM EDITION

The Invisible Fire

The Story of Mankind's Victory Over the Ancient Scourge of Smallpox

AN AUTHORS GUILD BACKINPRINT.COM EDITION

Published by iUniverse.com, Inc.

For information address:
iUniverse.com, Inc.
5220 S 16th, Ste. 200
Lincoln, NE 68512
www.iuniverse.com

Originally published by Putnam

ISBN: 0-595-16867-1

Printed in the United States of America

Acknowledgments

With considerable trepidation I acknowledge my debts to a large number of people who helped create this book. I know I'm going to forget someone.

Therefore I must begin these acknowledgments with an apology to whomever I have forgotten and hope that he or she understands it is far more likely the result of creeping senility than of ingratitude.

The book falls approximately into two sections. The first, the historical section extending through Chapter 8, concerns smallpox as it affected history, beginning with what might have been the first recorded epidemic, and running until the recent years when the world got together to do something that had never been done before—eradicate a disease. The second section—the journalism—recounts how that was done.

The historical section could not have been written without the help of a number of people and institutions. It is a fact that if I had lived in almost any other city except Philadelphia the book might not have been written. There are few places in America with its library and historical resources.

My main source was the College of Physicians of Philadelphia, America's oldest medical society, librarian Elliot Morse and his wizard staff. The library and museum of the college provided an unending supply of surprises, so most of the historical material you are about to read came from primary sources, including original letters handwritten by Thomas Jefferson, books hundreds of years old in perfect condition, pamphlets and manuscripts beyond price. The collection even includes a lock of Edward Jenner's hair, which, I should report, is brown.

The deepest bow goes to the college's former curator of historical books, Ellen Gartrell, now of the University of Pennsylvania. The Gartrell Quick Course in medical historical research should be bottled and sold, and she is a marvel. Her successor, Christine Ruggere, is right in the same league.

I am also indebted to the staff of the Library Company of Philadelphia (a historical library founded by Benjamin Franklin), to the Pennsylvania Historical Society and to the Free Library of Philadelphia. The libraries at Thomas Jefferson University, Temple University and the University of Pennsylvania also contributed to this work.

John Ashbrook, a colleague at the Philadelphia *Inquirer,* came up with the diary of his great-great-great-something-or-other, Lewis Beebe, the physician on the Quebec expedition. It had been published only once before, many years ago, and without John's quick thinking, I might not have known about it.

My debts for the journalism section are even longer. I am particularly indebted to the Center for Disease Control (CDC) in Atlanta, the foreign ministry of the government of

India, the government of the Somali Democratic Republic and, most of all, the World Health Organization. If there is any single hero to this book, it is an institution, WHO. It was WHO which eradicated smallpox.

Some of the arrangements for interviews and research were made through information officers, and four of them must be mentioned because they were of great assistance to me: Bob Alden of CDC and Joan Bush, Peter Ozorio and Jim Magee of WHO. I worked with Peter for several years on this project and I am particularly grateful to him.

I have lost track of all the people I interviewed for this book through the years, but I must mention some who were particularly helpful.

Atlanta: Walt Orenstein, Mary Guinan, Martha Thieme, John Obijeski, Rafe Henderson, Mike Lane, Stan Foster, David Sencer and Bill Foege. A special thanks goes to James Nakano of the smallpox lab for letting me visit his frightening little workshop and come face-to-face with one of the deadliest creatures on earth.

Geneva: Isao Arita, John Wickett, Alan Schnur, Victor Ladnyi, A. Monnier, Andy Agle, Nedd Willard and Nicole Grasset.

New Delhi: Pierre Ziegler, George Stroh, Himachal Som, M. I. D. Sharma, Jitendra Tuli, R. N. Basu and about a dozen other people in the Nirman Bhavan.

Lucknow: R. S. Bajpai, R. P. Singhal, Bagamber Singh and the medical staff of the Bakshi-ka-talab Primary Health Centre.

Mogadishu: Abdullahi Deria and Zdeno Jezek.

Merka: Ali Maow Maalin and Hassan Alasso Nur.

By telephone: Benjamin Blood of the National Institutes of Health; Benjamin Rubin and Malcolm Bierly of Wyeth Laboratories, Radnor, Pennsylvania.

Three people deserve special mention, for they symbolize the remarkable group that achieved the remarkable victory.

They are charter members of the "Order of the Bifurcated Needle." D. A. Henderson, dean of the Johns Hopkins University School of Public Health, gave me many hours of his valuable time in his Baltimore office, his home and on the telephone. I spent a fascinating and stimulating two days in Chelsea, Michigan, with Larry and Girija Brilliant and many hours on the telephone. I hope this book does them justice.

Any mistakes or misrepresentations found in this book are entirely my responsibility and could not possibly reflect on the memories or talents of any of the above-mentioned.

I must also thank some people who were very kind during my travels with a drink, a meal or a bed. I am indebted to D. A. and Nana Henderson in Baltimore, Gail and John Wickett in Geneva, Dr. and Mrs. M.I.D. Sharma in Delhi, the Ozorios of Fermé-Voltaire, Heidi Blattmann in Zurich and most especially Bonnie and Larry Pike in Atlanta. One of the pleasures of doing this book was in renewing my old friendship with the Pikes.

Pennie Marcus was invaluable in helping me with the research in this country. Martha Brannigan assisted overseas.

Eileen O'Brien edited the historical section with great skill and hilarity, although she reports coming down with a rash and fever during the gory parts. Mary Coady edited the modern section.

Ceil Coady typed the manuscript and processed the photographs.

The genesis of the book was an assignment by *Inquirer* national editor Steve Seplow, now of Knight-Ridder's Washington bureau. His successor, Jim Naughton, also appreciated the importance of what was occurring and helped get me overseas. Gene Roberts, the executive editor, kindly gave me the time off to pursue my research.

Dave Preston of Geneva was very kind in letting me

rummage through his files. I never met him but I owe him at least a beer.

My agent, Mike Hamilburg, gentleman and scholar, sometimes had more faith in this project than I had, and to him, perhaps, I owe my greatest debt. He is obviously the Hollywood agent Fred Allen never met.

My editor, Tom Hyman, was another whose faith was unwavering. That kind of support is necessary if one is going to take on a project of this size.

I must also thank Motria, Matty and Nicky for putting up with me.

And finally, to Lorna, for whom the ordeal must have seemed unbearable at times, thank you.

For Jonathan and Michael

Contents

Introduction

"That disease ... was then the most terrible of all the ministers of death. The havoc of the Plague had been far more rapid: but Plague had visited our shores only once or twice within living memory; and the small pox was always present, filling the churchyard with corpses, tormenting with constant fears all whom it had not yet stricken, leaving on those whose lives it spared the hideous traces of its power, turning the babe into a changeling at which the mother shuddered, and making the eyes and cheeks of a betrothed maiden objects of horror to the lover."

—Thomas Babington Macaulay,
History of England

"In expressing our triumph in the conquest, which reason and humanity have lately obtained over the most mortal of all diseases, we are led to anticipate the time, when they shall both obtain a similar conquest, by applying the means which have been discovered, to the prevention and annihilation of the plague. How long blindness, which has perpetuated this disease for so many ages may continue, I know not; but I believe it is as much out of the power of prejudice, error and interest, to prevent its final and total extinction, as it is to prevent the change of the seasons, or the annual revolutions of our globe around the sun. . . ."

—BENJAMIN RUSH in a letter to BENJAMIN WATERHOUSE, February 9, 1801

"The *I Ching,* the oldest book in the world, suggests two ways of organizing people to accomplish great tasks. One way is that of an army; warlike, with strict obedience within and a central person to which all others pay obeisance. The second way the *Ching* calls "Fellowship of Men" (T'ung Jen), a gathering together not because of one hard, strong personality which dominates all others, and not because of some private individual interests. True fellowship among men is based upon a concern that is universal, upon the goals of humanity. By sacrifice and sharing people come together and in their unity are inspired to ever greater heights. . . ."

—LARRY BRILLIANT, M.D.
July 22, 1978

This is the story of one fellowship which did something that had never been done before in all of human history: completely, totally eradicate a killer disease, one of mankind's greatest scourges. In doing this they reached heights of service beyond recompense.

I

The Elegant Killer

James Nakano stripped off his clothing, put his watch and rings into his pants pocket and hung his things on the wooden rack screwed into the gray-tiled walls.

The dressing room, accessible only to the few researchers granted a key, was small and cluttered. Besides the clothes rack, there was a large white waste receptacle, a sink and mirror, and a small locker. Many of the men who could use the room had left some article of clothing on the rack, so it was nearly full even before Nakano added his street clothes to the collection. Above the rack were freshly laundered white towels, some of them looking as if they had been used as bath mats and still bearing not completely washable footprints.

Nakano, now wearing only his glasses (a rare exception to the security regulations), glanced behind him to make sure the outer door was closed, and then walked through the small, silent shower room into the eerie, glowing, blue-purple room beyond it. He had left the "dirty area" and was now entering the "clean area." It was not what Nakano was going to take into the laboratory that necessitated all the care but what he might take out.

The "clean room" was lit by ultraviolet lights strung on a laundry rack festooned with surgical clothing. Only a bit of normal light crept over the shower curtain behind him to soften the garish deep blue of the ultraviolets. The rest of the room was lost in deep purple and, finally, total darkness.

Nakano took the gray-green clothing from the rack: a short-sleeved, deep-necked shirt and a pair of pants with a cloth belt. Then he raised his arms over his head to a wooden compartment, swung up the door and reached inside. Another array of ultraviolet lights from the compartment now glared at him.

The glare distorted his pleasant, round face. A Californian of Japanese ancestry, Nakano was educated in Britain and is now in charge of the care, feeding and breeding of some of the deadliest particles ever to inhabit this planet. All the safety procedures are taken seriously, because Nakano knows what might happen if they are not.

He remained intent as he found the rubber tennis shoes with the initials inked on the toes, took them out and closed the compartment door. His face was plunged back into the purple gloom.

Behind the wall to his right was a room for the women researchers just like this one. In all only eighteen people are ever allowed to get this far into the laboratory.

Nakano opened the final door and entered one of the most dangerous rooms in the world.

The lab in which he was now standing was located on the fourth floor of Building 7 of the Center for Disease Control in the quiet Druid Hills suburb of Atlanta. The building is one of a cluster of interconnected red-and-tan structures located next to the Georgian-style marble campus of Emory University.

From CDC go some of the world's best medical detectives to track down men's killers, common or rare, frequently stalking them to some of the remotest places in the world. Back come samples for study, many of them containing deadly viruses and bacteria. It is one of the most dangerous jobs in medicine to be an epidemiologist. The number of scientists who have come down with the disease they were studying—sometimes with fatal results—is shocking. The job can also be one of the most rewarding in science. It was CDC that produced many of the people and the winning strategy for the eradication of one of the greatest slaughterers of mankind, the virus variola.

A small sign on the door outside the fourth floor walkway tells the story of Nakano's lab:

BIOLOGICAL HAZARD

DO NOT ENTER

SMALLPOX

Nakano flicked on the fluorescent lights. When the lab is idle the only illumination comes from two rows of ultraviolet lights over the two work tables that jut out from the near wall. No one can work under ultraviolet lights because of the poor visibility, the danger to the eyes, and the severe skin burn that would result from the radiation, so when work is being done normal lighting is used and the ultraviolets are shut off.

The ultraviolets are essential to the safety of the lab. It is one of the many security devices designed to make sure that no smallpox ever leaves the laboratory "clean" rooms. Ultraviolet radiation will kill any viruses that pass within a few inches of the bulbs. In the "clean" room the clothing and lab shoes are constantly bathed in an ultraviolet glare, and in the lab itself the air is constantly circulated so that sooner or later any free viruses left behind by the scientists will be carried past an ultraviolet bulb and be destroyed.

If that doesn't happen, the lab's filtering system will catch them. All the air leaving the lab is passed through a series of filters, each containing holes much smaller than the known size of pox viruses.

Nakano says this guarantees with 99.95 percent certainty that no virus leaves the room through the air-conditioning system.

As an additional precaution, the air pressure inside the lab is kept lower than the normal atmosphere outside the room. If there is any leak in the airtight system, air would rush into the lab from the outside and no deadly virus would be able to leak out. If any of the seals of the lab are broken, a red light alongside the main door goes on, a signal that security has been breached. The complex system of doors acts like an air lock in a spaceship to protect the integrity of the system.

Alongside the entranceway to the lab Nakano stopped beside three gray freezers: two of them chest models, the other an upright. All three were stoutly padlocked.

Inside the freezers on metal racks are glass vials containing samples either taken from smallpox victims or viruses bred in the lab for purity and virulence. There are enough viruses in those freezers to destroy about half of the unprotected population of the United States in weeks. Since few people have currently effective smallpox vaccinations these days, that means most of us.

The viruses are kept frozen—safe and docile—at a temperature of −70° C (−94° F). When work is being done with a sample, the viruses are taken out and placed on the gray slate work tables to thaw for about ten minutes.

Much of the more dangerous work is done at one of the two security chests along the wall. They are sealed, glass-fronted cabinets, each fitted with four portholes through which the researchers can stick their arms. The arms slide into black rubber sleeves attached to the portholes, which end in thick yellow gloves. Samples are placed in an air lock between the two chambers. Only when the outer door has been closed can the inside door be opened with the gloves. Much of the work in breeding new strains to keep the laboratory equipped with high-potency viruses is done through the gloves in these chambers.

Besides having their own negative pressure, each chamber is also connected on its own to the main filtering system.

Work in which there is less danger of contamination is done on the work tables that make the lab look like nothing more complex than a college biology laboratory, with glass vials, Bunsen burners and small microscopes littered about.

It has a deceptive look of innocence, this lab. Yet the ghastly death is there, silent, invisible, odorless—perhaps sitting on the slate table or floating near your head on an unfelt wisp of air, or on the lip of a test tube, on the handle of the freezer, in tiny cuts in the rubber gloves. Probably more people have succumbed to it than to any other plague in history. More certainly have been disfigured or blinded by it. It remains incurable, yet because of the work of people like Nakano and labs like this one, it may never strike again.

CDC is determined, of course, that if it does strike it won't be because of a leak from the laboratory, so the process Nakano will use to get out of the lab and take samples for study is even more elaborate than the process for getting in. No object is permitted to leave the room unless it has been either autoclaved or soaked with ultraviolet radiation. The sample Nakano is taking out for study in the electron microscope must first be killed very dead with formaldehyde or Clorox. The plastic containers holding it come in baked with ultraviolet radiation.

Other equipment is placed in sealed plastic bags. The bags are then washed in Clorox and with their contents are placed in an air lock next to an ultraviolet bulb for at least half an hour.

Once when a television crew filmed inside the lab a special hole in the wall had to be constructed so that the cameras could be bagged, soaked and zapped with ultraviolets.

Humans cannot be autoclaved or soaked in formaldehyde, so the best that can be done for them is a thorough cleaning.

There is really no such thing as a smallpox carrier in the sense that you can pass on the disease without having it. Lab workers and others who have an immunity through vaccination or a previous infection may be able to carry some virus in their throats or nostrils, but the chance that they can pass it on to someone else is remote, Nakano says. The virus would not be able to multiply as it would in an unprotected person because the antibodies would kill the virus first. Everyone working in Nakano's lab—or even entering Building 7—must have been vaccinated within a year.

Theoretically all it takes is one individual virus organism to touch off a case of smallpox and start an epidemic, but, in reality, that never happens. Except in a laboratory, no one is likely to come into contact with just one smallpox virus. The viruses go in gangs of millions.

"There is a possibility that you could carry one, two or three particles from the lab, but I doubt very much that you could start anything," Nakano says. "The possibility of people carrying smallpox is very rare."

Even so, CDC takes no chances that the disease will leave the "clean" areas on the bodies or clothing of the researchers.

After Nakano placed his sample under the ultraviolet light in the air lock at the far end of the laboratory, he returned to the "clean area" dressing room, carefully locking the door behind him. The lab clothing was placed on the ultraviolet laundry rack and his shoes were put back into the wooden compartment under those bulbs.

Nakano then stepped into the one-man shower room that separates the two areas and bathed. Special soaps are not necessary, but Nakano was careful to wash the areas exposed by the somewhat informal surgical garb. Since he wore a cap he did not have to wash his hair, but he took off his glasses and soaped and soaked them.

Finally, he blew his nose into his hands to clear anything that might be hiding in his nasal passages. Then, and only then, could he go into the "dirty area" dressing room and get his street clothing.

The invisibility of viruses makes them hard to deal with because you never know when they are around until it is too late. Besides being invisible they are also exceedingly efficient parasites, doing their dirty work in a frighteningly direct, uncompromising manner.

The only way you can actually see a virus is with an electron microscope. Enlargements of 100,000 times are necessary for routine microbiology, and only these machines can do it.

One of the researchers at CDC who can handle the machinery is Martha Thieme, who was waiting for Nakano and his sample in the basement of Building 7. She is on a special list of scientists on twenty-four-hour call; any time a World Health Organization field team thinks it may have a smallpox case a specimen is flown to Atlanta for a quick identification.

In the spring of 1976, just as WHO believed it had finally licked the disease, new suspected cases were reported in Ethiopia. Samples were flown to CDC, and it was Thieme and her co-workers who verified the bad news: it was smallpox.

Similarly, when suspicious cases later popped up in Somalia among nomads who frequently crossed the remote border with Ethiopia, it was the scientists at CDC who confirmed with their electron microscopes that the disease had spread, again delaying the historic announcement that smallpox had been eradicated.

The sample Nakano handed Martha Thieme was from a Somali named Olad Hajji Yussof. A WHO physician had swabbed a sore on Yussof's arm. By that time, the fact that there was smallpox in Mogadishu, Somalia's capital, had been confirmed, so everyone knew what to expect. Yussof was—it was hoped—going to be one of the last cases—ever.

The electron microscope, a tubular device flanked by two control panels, sat on a table in a small room. Thieme asked Nakano to turn out the lights as she began to flood the magnification chamber with electrons.

The sample was contained on a tiny square copper screen with the material spread across the mesh. Thieme removed it with tweezers and inserted it into a section of a tool that looks a bit like a hypodermic syringe. She inserted the tool into the base of the microscope.

At the bottom of the machine were three windows opening up into a chamber. One could see clearly the wire screen enlarged gigantically on the glowing floor of the chamber, each tiny wire now the size of a thick rope as the magnification field studied each cross section.

Inside the chamber everything was lit in a pale green light. Interestingly, neither Thieme nor Nakano could see the color of the light; both are color-blind and see the inside of the chamber only in black and white. Their color blindness probably helps them see in the electron microscope chamber with better contrast than a normally sighted person. (Nakano also points out that he has an unusual number of left-handed persons in his department but has no idea what that means.)

Blurs and scratches seemed to appear between the shadowy wires on the floor of the magnification chamber. A spoon-shaped shadow flitted briefly into view as Thieme moved the field of vision, adjusted the focus by rotating a cylinder on each side of the microscope and changed the magnification with a dial to her right.

When she found what she was looking for she let Nakano look into the twin eyepieces.

What he saw did not require color blindness or special vision. It was quite clear. In the middle of the screen was a large shadow that built and unfolded as the depth of field was changed. In three dimensions, it looked like a brick. Nakano could not tell which strain of smallpox virus he was looking at, but it was surely smallpox.

It was dead now, but it must have given Yussof a very hard time just a few days before. Next to the virus lay the remains of a slaughtered cell. The virus he was looking at was probably one of the unwanted children of that cell—and one of its murderers.

The disease is almost unprecedented in its ghastliness.

The incubation period—the time from the first assault to the first symptoms—is about twelve days in classic smallpox. The onset is sudden and dire.

One minute a person is walking about, apparently healthy, thinking those thoughts all of us think when we can ignore our bodies, watching the world, participating in it.

Suddenly he is stricken and his world is submerged in pain. There is a splitting headache and a knifing sensation in the back. He gets chills and fever and often that terrible malaise—the feeling that his body is suddenly out of gear and will no longer work. If the victim is a child convulsions are possible.

The second day is worse than the first. The patient is now flat on his back, his temperature up to 104° F (40° C), clinging to his world by his fingertips, wishing he could die quickly. He cannot. He becomes delirious and slips gradually into a stupor, and sometimes even a coma.

On the third or fourth day of the attack the symptoms abate. The temperature sinks to near normal and the mental clouds seem to roll away. The victim now looks and feels much better, and unless he knows what the disease is, he may actually think the worst is over. There is, after all, only a rash spreading on his face and a little soreness in his throat. It is easy to assume that nothing calamitous will happen now.

The first pox sores appear in the mouth and throat. They become very painful very quickly, feeling something like a mouth full of canker sores and orange juice. The throat swells. The patient gets hoarser.

Sores now appear on the face and forearms, spreading to the upper arms and trunk, congregating particularly on the back. Finally they reach the legs. The sores go through an evolution into what doctors call macules, papules, vesicles and, finally, pustules. That means that the sores appear first as red splotches which swell, spread and erupt into hideous wounds that eventually scab over. By the time it is done growing each sore is about eight millimeters (0.3 inch) in diameter.

As the eruptions begin the temperature spikes upward again, generally passing 103° F (39.4° C). One textbook, in classic scientific understatement, notes that "at this time the patient presents a dramatic picture of a miserable person." Indeed. The sores in his mouth and throat are excruciating. The face is swollen beyond recognition and is marred by the scalding pockets of pus.

While all this is happening the eyes are under attack. Many cases end in permanent blindness, although the exact mechanism is unknown. At its height, smallpox was one of the leading causes of blindness in the world.

There are secondary infections from bacteriologic sources; some people die from these, not the smallpox. Exactly what causes the death in uncomplicated smallpox is unknown.

The disease runs its course in about two weeks. It takes a month for the last scab to fall off. If the victim survives he will probably be scarred for life. He has gone through one of the most horrible ordeals imaginable.

During the course of the disease his physician—if the patient has one—can get a fairly good idea of his chances of survival from the symptoms. If the sores do not touch (discrete smallpox), he is a lucky man. Less than 10 percent of such cases end in death. If the sores are so close that they run into each other to form huge pustules (confluent smallpox) the mortality rate is about 50 percent. If bleeding erupts beneath the surface of the skin and from the nose and mouth (hemorrhagic smallpox) death is certain. No one survives.

There are instances in the literature in which hemorrhagic smallpox is called purpura variolosa or black smallpox.

Another clue to survivability is how badly the victim feels during the pre-rash days. The greater the discomfort the likelier that the case will be serious. Age is also important; the death rate is much higher for the very young and the very old. Gender makes no apparent difference but prior vaccination does. Those who were vaccinated at least once generally fare better than the never-vaccinated; the more recent the vaccination, the better.

A study of 1,318 patients in a Calcutta hospital during 1973 showed not only that old vaccinations reduce the "attack rate," but also that they mitigate the severity of the disease and the fatality rate. Persons vaccinated just before infection were unlikely to contract the disease in any form.

Vaccination after infection seems to help somewhat. In the Indian study the overall death rate was 53.4 percent.

Among those vaccinated within the incubation period but before the onset of the rash it was 44 percent. Once the fever broke out vaccination did not help.

During the height of all this terror the victim is spreading the virus into his environment, spreading the misery. He is most infectious during the first two weeks, breathing out millions of viral particles, mostly in the form of droplets that float in the air on every breeze, clinging to utensils, clothing, sheets, blankets, dressings. Infected scabs are frequently mentioned as being notorious for spreading the disease, but there is little corroborating data.

Although, as we shall see, throughout history hundreds have proclaimed "cures" for smallpox, none of them really worked. There is no cure or treatment capable of stemming the tide of infection. Treatment is symptomatic, which means that the doctor tries to ease the suffering while knowing that he can't do anything about the outcome of the smallpox attack as such. Antibiotics, usually penicillin or tetracycline, are administered to prevent secondary infections. Because of the painful ulcers in the mouth and throat the patient is fed intravenously. Precious bodily fluids and chemicals are replaced in the same way to prevent dehydration and deficiency complications. Aspirin or codeine are used to help fight the head and back pain, and there are some experimental drugs that may help prevent blindness, although generally all that can be done is to keep the eyes clean. Most hemorrhagic patients must be treated for shock.

Once the disease has begun to spread through a population, there are a few procedures besides immediate wide-scale vaccination that seem to help. One is to make a serum from the blood of an immune person and inject it into someone who has come into contact with the disease a day or so after being vaccinated.

An antiviral agent, n-methylisatin-beta-thiosemicarbazone (or Marboran), has been found to be effective in preventing smallpox in some contacts, although no one knows exactly why it works, and the data are not clear.

That is the modern treatment response to a smallpox epidemic. Most victims throughout history and throughout the eradication programs never saw n-methylisatin-beta-thiosemicarbazone. They suffered unassisted. There is little evidence from recent times that such sufferers experienced a markedly higher death rate than those who received the treatments described. At least victims in the twentieth century were far better off than those in the hands of earlier physicians.

Throughout history the best way to stem an epidemic was by quarantine. Usually special hospitals were built for the express purpose of isolating smallpox cases, and some of those still exist today. Sometimes they were located on remote islands, or in desolate buildings, and most flew flags to warn away the unwary. But sometimes the hospitals wound up spreading the disease instead of preventing its spread.

Hospital isolation was still the rule until the recent eradication. Victims were kept away from nonimmune people as long as there was a single crust left on their bodies. Every bodily discharge and all their waste material were placed in paper bags for incineration. All their utensils and dressings were sterilized by steam or boiling. Every thing was scrubbed and sterile.

All of this to prevent something from getting loose.

The knowledge that there are such things as viruses which can cause diseases like smallpox came about almost by a process of elimination.

Since antiquity unseen germs were suspected of causing illness. The Roman, Marcus Terentius Varro, in the first century B.C. wrote: "If there are any marshy places, little animals multiply which the eye cannot discern, but which enter the body through the mouth and nose and cause grave diseases." Girolamo Fracastoro, in 1546, made much the same suggestion, adding that the little creatures probably could reproduce themselves.

The first sighting of a one-celled microorganism was recorded in 1683 by the Dutchman Anton van Leeuwenhoek, who came face-to-face through his microscope with a bacterium. He did not know what it was, however, and made no association between what he saw and diseases.

The first man to sense a connection apparently was the Italian Agostino Bassi who, in 1835, found bacteria on the body of silkworms suffering from a disease called *calcinaccio.* Louis Pasteur is credited with finally proving the germ theory, although he too was dealing mainly with bacteria.

To help scientists isolate bacteria, a contemporary of Pasteur, Charles Chamberland, developed a handy tool, a series of porcelain filters in the shape of a candle that filtered bacteria from sample liquids. The problem with Chamberland candles was that they didn't succeed in isolating a suspected organism in every instance.

In 1892, for example, Dimitri Ivanovsky took fluid from a plant affected with tobacco mosaic disease and passed it through his Chamberland candle to eliminate the causative organism. He then spread the filtered fluid on healthy plants and found that they came down with the disease. Something had managed to get through with the liquid and Ivanovsky could only conclude that he had poor filters.

In 1898 the Dutchman Martinus Willem Beijerinck, conducting the same experiment, encountered the same result. He reasoned, however, that the liquid must have contained an organism, much smaller than any known bacterium, that could pass through the filters. He even extrapolated on its nature, concluding that probably it did not behave much like bacteria and was totally different in habit.

He was exactly right. Tobacco mosaic disease is caused by a virus.

Scientists love to debate whether viruses are alive or not.

They have a set definition for living things. If an organism fits that description scientists can easily define it as "living." If it does not conform, it is not "alive."

One of the things conforming living organisms do is reproduce either sexually or asexually. Viruses do not conform to this standard. They neither fornicate nor split. They reproduce by gang attacks on cells, a form of reproduction by murder.

Viruses are devilishly simple creatures, having a beautiful simplicity that scientists describe as "elegant."

The most distinctive difference between viruses and most other living things concerns the chemical compounds called nucleic acids. Deoxyribonucleic acid (DNA) and ribonucleic acid (RNA) are of principal importance. Unlike most organisms, which have both DNA and RNA, viruses have only one or the other.

DNA is the master of genetics. Like a computer tape programmed to perform a certain complex function, DNA sits in the nucleus of a cell and gives the orders for heredity. If the cell will grow into a creature with blue eyes there is a protein in the cell's DNA containing the order for blue eye color. If the creature will have dark hair, the DNA is programmed to order it.

RNA is the good soldier who slavishly follows the orders of DNA. The two substances chemically are about the same, but they have different sugar compounds.

The viruses containing only RNA are called, logically enough, RNA viruses. Polio virus is one of these. The pox viruses fall within the category of DNA viruses.

And unlike most living things, viruses do not carry around spare baggage. They contain only the DNA or RNA necessary to hijack a cell and force it to produce more viruses.

They care not for blue eyes and dark hair. They carry only the equipment they need to commit murder and multiply. They are very elegant killers.

There are four viruses associated with the disease we call

smallpox. The three main ones are called variola viruses. One is variola major, which is the real killer, the classic smallpox. It destroys from 15 to 40 percent of its victims, blinds many and disfigures nearly all. Another is variola minor, a virus that causes pox that disfigures many of its victims but kills very few. Between the two is a whole spectrum of strains called collectively variola intermedia, sometimes called the African strain. It has a kill ratio of about 5 to 15 percent.

The fourth virus involved in this tale is vaccinia. It is the virus used in smallpox vaccination. The odd thing about vaccinia is that it does not exist in nature and probably did not exist at all before Jenner began using cowpox for his smallpox vaccine. Probably it was mutated from a cowpox virus sometime in the last 200 years but no one knows how or why or when. Suddenly scientists found that the vaccine which they were using contained a virus they had never seen before and which was different from the one they thought they were using. It is one of the great mysteries of medical history, and it is doubtful that anyone will ever find out who, if anyone, did the deed, or if nature altered the virus unassisted.

The four viruses are part of a much larger family called orthopox viruses. Most are classified by the host they are known to infect, so there are camelpox viruses, raccoonpox viruses, monkeypox viruses and, unaccountably, one recorded case of "Lenny" pox viruses (which, perhaps, infected someone named Lenny).

There are a total of fourteen orthopox viruses. They are related to six other groups, a total of forty-six viruses in all that produce pox diseases in humans, animals or in both. The only things all forty-six seem to have in common, aside from the type of disease they cause, are a single protein with a molecular weight of about 70, and their size and structure.

Smallpox, vaccinia and "Lenny" seem to work in humans

only. Monkeypox infects both humans and other primates. There have been thirty-six recorded cases of monkeypox in humans, five of them ending in death, although it is not known if the cause of death was the pox or the victims' living conditions.

Molluscum contagiosum (which produces benign wartlike tumors), tanapox and yaba also infect both humans and other primates. Some viruses, like cowpox, infect humans and nonprimates. And many more infect only animals and not, to the best of our knowledge, humans. It is hoped none will alter its life-style and invade people—a danger smallpox fighters fear.

Scientists are quick to point out that although viruses look very simple in electron microscopes and in line drawings, they can be deceptively complex, especially pox viruses.

The variola and vaccinia viruses have something of a nucleus, a dumbbell-shaped structure in the center filled with the killer DNA. There are two lateral bodies with still unknown functions on each side of the dumbbell.

The virus measures about 300 millimicrons (or nanometers) long, which is about 12/1,000,000 inch (3/1,000,000 centimeter). Smaller viruses run about 10 millimicrons.

Pox viruses are the largest known viruses and pack a great deal of equipment. In fact, they're also the most complicated viruses known. The genetic information in the nucleoid weighs about 200 million Daltons (a unit of atomic weight). Polio viruses, which are about average in size, have genetic material weighing only about 1.6 million Daltons. As CDC biochemist John F. Obijeski describes it, variola is about 100 times more likely to kill you. In fact the DNA in the variola and vaccinia viruses form the classical double helix shape found in more complex nonviral organisms.

Vaccinia is the biggest of all the viruses. It is almost large enough to be visible under optical microscopes. When it clusters in specimens, which viruses are wont to do from time to time, it can be seen without an electron microscope.

In order to tell the three variolas and the vaccinia apart scientists have to use genetic marking systems to pick out the DNA structure, but the electron microscope serves as a perfect tool for identifying a disease which is frequently misdiagnosed in its early stages. For instance, not even experts in the field can be sure if they are looking at an early case of smallpox or at one of chickenpox. But under an electron microscope the difference can be seen easily in specimens. Chickenpox is caused by a virus of the herpes family, which is much smaller and looks nothing like variola or vaccinia.

The structure of the pox virus is now fairly well known. Surrounding the nucleoid of variola and vaccinia is an outer coating of protein, cholesterol and fat, about 20 to 30 millimicrons thick.

In contrast to polio viruses, which have 4 proteins in their structures, pox viruses have 30 to 35. That does not count the proteins in the DNA.

Viruses tend to be somewhat adaptive. Strains of variola grown in laboratories through generations tend to lose a protein or two. The virus apparently decides it doesn't need all those functions and, being naturally parsimonious, drops the unnecessary ones.

Viruses do lose some potency after a few years of inactivity in laboratories, which is why Nakano and his staffers must breed some strains in egg cultures to maintain the virulence of their test samples.

Even among viruses, variola is unique for its clean, murderous ability.

The variola viruses generally enter the body through the nose, having been exhaled by another victim. They always originate from another human.

Unlike many other viruses, the life cycle of variola goes only from human to human. There is no animal reservoir, no place it can hide when it runs out of victims. This is crucial in understanding how the disease was wiped out. It is the one flaw in an otherwise perfect killer.

The first contact a virus has, in most cases, is with a cell in the mucous membrane of the lungs. Just how the assault takes place is not entirely clear. Somehow the protein shield of the variola becomes attached, or adsorbed, as scientists call it, to the outer wall of the cell. Some viruses are very choosy about the way they are adsorbed and dictate tight rules for the process. Variola doesn't care for niceties. It will take adsorption any way it can find it.

The next step occurs when the cell makes a terrible mistake. It engulfs the variola in a fatal embrace.

As soon as the virus has been engulfed, its outer shield drops away, exposing the deadly interior, the DNA.

In order to perpetuate itself a virus has to do two things. It must make more DNA to pass on the genetic information to its progeny, and it must produce proteins for the viral particles about to be born.

So the first thing that happens in the victimized cell is that the DNA from the virus synthesizes messenger RNA, which is the chemical imprint of its orders. One of the most important functions of messenger RNA is the production of the chemical enzyme that makes more DNA. Since the virus did not bother to carry any of the raw ingredients for the job, this enzyme forcibly takes material from the captured cell. It is a kind of chemical rape.

Another type of messenger RNA is busy synthesizing the protein not only for the new viruses but for the enzymes that power the reaction now taking place in the cell.

Looking through an electron microscope, a scientist can follow the action as a shadow appears in the cytoplasm of the infected cell, the portion that surrounds the nucleus. Here the dirty work will be done. Most other viruses go straight for the cell's nucleus to reproduce, but not pox viruses. They need lots of room.

Gradually these darkened areas seem to grow, get very dense and then produce membranes at the outer edge. Scientists call these areas "factories," and the term is very apt.

In a matter of a few hours these factories become a nest of viruses, exact copies of the original assailant, not just a few, but anywhere between 10,000 to 100,000 viruses crowding in this one poor cell, where to start there was but one.

And unlike other viruses, which kill the cell the second the reproduction has been completed, deflating it like a balloon so that the particles can spread themselves through the body, variola colonies keep their cells alive and about 90 percent of all the new viruses stay in the adopted home.

The disease is spread by cell-to-cell contact. The cells in the lungs are stacked one on top of the other, and an infected cell easily spreads variola to its neighbors, which then undergo the same indignity.

Sometimes the cell will get into the bloodstream and, as it goes bumping through the circulatory system, trail death behind it. Eventually, of course, the cell will die and the viruses still inside will decide to leave and find another host.

The death of the cell also sets loose a number of chemicals to spread secondary poisons in the body, where they do not belong.

By now, however, the body has sensed the intrusion and calls on its defenses.

The first line of defense is to prevent the viruses from ever reaching the internal cells at all. The skin provides impenetrable armor against disease-causing organisms so long as it stays unbroken. The bodily orifices, like the nose and mouth, are also designed to repel invaders. The nasal passages contain tiny waving arms—cilia—which cleanse the air of alien life forms. Mucous membranes are constantly being washed to prevent germs from affixing themselves. Coughing and sneezing are other ways the body protects itself. Should an invader manage to get into the stomach it runs into hydrochloric acid, which not only aids digestion but is quite capable of killing disease bearers.

There is also an early-warning system called interferon. A cell attacked by a virus releases a substance that runs riot, screaming to other cells in the area that there is an invader loose. This Paul Revere of the body does not actually attack the virus, but it seems to stimulate the surrounding cells to produce a protein that makes it impossible for the virus to reproduce inside them. No one knows exactly how this inhibiting process works (interferon was not discovered until 1957), but there are hints that it short-circuits the messenger RNA produced by the virus so as to make it impossible for the RNA to follow its orders. Interferon is now believed to be the major defense system bodies have against viruses.

But obviously this elaborate defensive package fails in the face of many assaults. If it did not we would never get sick.

After the initial infection, the viruses spread through the body. Most head for the epithelial layer of the skin, their favorite playground, but many also congregate in the throat and upper respiratory tract. Should the victim die of a variola attack, an autopsy will probably reveal that the liver, spleen, sexual organs and other parts of the body have been under attack and that the virus left almost no organ untouched in its efforts to produce a ghastly death.

Once the invasion is under way, the body must call on more complex defenses.

One defensive weapon is the phagocytes, cells found in the blood, lymph nodes and tissues. Phagocytes engulf and eat viruses and bacteria. One reason why lymph nodes, like those in the armpit and neck, swell as a sign of infection, is that the body is sending its army of phagocytes into the lymphatic system to counter an attack and they are massing in the nodes to join battle. If the viruses manage to get out of the lymphatic system alive, they are instantly sniped at by other phagocytes in the bloodstream or in surrounding tissues.

To assist the phagocytes the body frequently pours extra blood into the infected area, in effect carrying even more troops to the battle. There is some coagulation to isolate that battle zone and an increase in temperature, resulting in what we know as inflammation and fever.

The problem with this defensive system is that there are some viruses which love phagocytes, and instead of being digested, make themselves at home in their pursuers, reproducing and spreading within them—medical Trojan horses.

Another defensive system involves antibodies. Viruses are antigens, that is they trigger a chemical reaction in the body that produces tiny soldiers that surround the viruses and seal off their outer shell, making it difficult, if not impossible, for the virus to be adsorbed. The antibodies are produced in the lymphatic system, coming from a large number of cells capable of producing the particular kind of antibody for each antigen.

After a viral attack the antibodies stay in the system, gradually decreasing in number until, eventually, they disappear (although in some diseases enough remain to last a lifetime). As long as antibodies persist in the system, any new attack by the same or related virus triggers a quick, massive rebuilding of the antibody army. This produces fast immunity and explains how vaccines work, and why once someone has had certain viral diseases he never gets them again. The effectiveness not only against the specific disease but sometimes against diseases caused by similar viruses also explains why vaccinia confers immunity to variola and cowpox to smallpox.

Sometimes viruses are not that easy to destroy. Many have the capability of mutating sufficiently after a time to make an end run around the antibody system. Influenza virus is famous for rapid mutation, so getting flu this year generally will not prevent an infection ten years hence. The virus mutates to the point where the antibodies for this year's disease will not trigger protection against the altered virus.

As far as anyone knows variola does not mutate so readily. But it is "elegant" enough to beat back most of the protective systems the body enlists to fight infection, which makes it synonymous with terror.

In most parts of the world, word of a smallpox epidemic used to be the most terrifying news imaginable.

As we shall see there is good reason for the terror. Since almost time out of mind it has been both the scourge of peasants and the death of kings.

II

The Plague of Athens

The war opened for the season in the first days of the summer, 430 B.C. As it had the year before—and as it would for many years thereafter—the summer brought an invasion of Attica by the Peloponnesians and, in reaction, raids along the Peloponnesian coast by Athenian ships.

Again the invasion was led by Archidamus, the king of Sparta. Pericles, in firm command of the Athenians, had ordered all the people from the countryside into the city where they could be protected from the Spartans behind the long walls. The result was much grumbling by the citizens at the inconvenience and at the fact that the Spartans and their allies now had free run of the countryside.

The Athenians were sorely pressed but well protected behind the walls and, best of all, wrote their chronicler, Thucydides, son of Olorus, they were healthy.

"That year, then, is admitted to have been otherwise unprecedentedly free from sickness," he wrote.

Their good health would not last long, and before the summer was out the Athenians had something far greater to worry about than the Spartans beyond their walls.

According to the reports recorded by Thucydides, the plague first began in parts of Ethiopia and then struck Egypt and Libya. Late in the spring a ship from North Africa docked at Piraeus, Athens' port; on it was someone very ill. He was bringing to Greece a disease never known there before. As occurs whenever a disease hits a virgin population, it produced devastating effects. Rumors quickly flew through the city that the Spartans had poisoned the water supply.

"All speculation as to its origin and its causes, if causes can be found adequate to produce so great a disturbance, I leave to other writers, whether lay or professional," Thucydides wrote. "For myself, I shall simply set down its nature, and explain the symptoms by which perhaps it may be recognized by the student, if it should ever break out again. This I can do, as I had the disease myself, and watched its operation in the case of others."

Thucydides could get very little help from Athens' physicians because soon they themselves were dying in great numbers, having spent most of their time with the infected. Nothing the physicians could do would stem the epidemic spreading that summer, and "supplications to the temples, divinations, and so forth were found equally futile," Thucydides wrote, "till the overwhelming nature of the disaster at last put a stop to them altogether."

> . . . people in good health were all of a sudden attacked by violent heats in the head, and redness and inflammation in the eyes, the inward parts, such as the throat or tongue, becoming bloody and emitting an unnatural and fetid breath. These symptoms were fol-

> lowed by sneezing and hoarseness, after which the pain soon reached the chest, and produced a hard cough. When it fixed in the stomach, it upset it; and discharges of bile of every kind named by physicians ensued, accompanied by very great distress. In most cases also an ineffectual retching followed, producing violent spasms which in some cases ceased soon after, in others much later.
>
> Externally the body was not very hot to the touch, nor pale in its appearance, but reddish, livid and breaking out into small pustules and ulcers. But internally it burned so that the patient could not bear to have on him clothing or linen even of the lightest description; or indeed to be otherwise than stark naked. What they would have liked best would have been to throw themselves into cold water; as indeed was done by some of the neglected sick, who plunged into the rain tanks in their agonies of unquenchable thirst; though it made no difference whether they drank little or much. Besides this, the miserable feeling of not being able to rest or sleep never ceased to torment them.

The bodies did not appear to waste away as they did with other illnesses. Thucydides remarked that when death came, usually after the seventh or eighth day, the bodies still looked as if there were considerable strength left in them.

Those who survived the first week went on to greater agonies. The disease seemed to settle in the bowels, "inducing a violent ulceration there" and rampant diarrhea. The diarrhea so sapped the strength of the victims as to kill many.

The disease, which began in the head, soon spread throughout the body, Thucydides wrote, "and even where it did not prove mortal, it still left its mark on the extremities; for it settled in the privy parts, the fingers and the toes, and many escaped with the loss of these, some too with that of their eyes. Others again were seized with an entire loss of memory on their first recovery, and did not know either themselves or their friends."

No other ailment was reported in Athens during this time, as if the cause of this plague expelled all others. The city turned into a charnel house.

Whatever was done for the hapless victims seemed to make little difference. "Some died in neglect, others in the midst of every attention. No remedy was found that could be used as a specific," Thucydides wrote, "for what did good in one case, did harm in another."

Clearly the physicians—those who survived—had no idea how to cope with this new malady. There was no common denominator among those stricken. "Strong and weak constitutions proved equally incapable of resistance, all alike being swept away."

What terrified Thucydides most was the emotional disintegration that accompanied the plague. One who became ill immediately sank into a bottomless depression as if the world was about to come to an end. This despair sapped the powers of resistance and, Thucydides felt, contributed to the malady's ravages. It was hard to ignore a city full of men "dying like sheep" and particularly depressing when the disease was caught by nursing friends and relatives. Giving assistance—acting with honor—was the most dangerous course to follow that summer in Athens.

A tragic dilemma resulted. "On the one hand, if they were afraid to visit each other, they perished from neglect; indeed many houses were emptied of their inmates for want of a nurse. On the other, if they ventured to do so, death was the consequence."

Many acted with great bravery, Thucydides reported, sometimes entering a house to help the afflicted even after members of the family could no longer stand the screams and moans.

Fortunately it quickly became apparent that survivors of this plague became immune. Fearlessly, such survivors were the best nurses. They had acquired an illusion of general invulnerability. "In the elation of the moment [they] half entertained the vain hope that they were for the future safe from any disease whatsoever."

Pericles' stratagem of emptying out the countryside to protect the citizens now proved disastrous. Farmers and their families found no proper lodging in the crammed city and packed themselves into hovels and temples. They were bedeviled by the oppressive summer heat and particularly vulnerable to whatever was hunting humans in the city.

> The bodies of dying men lay one upon another, and half-dead creatures reeled about the streets and gathered round all the fountains in their longing for water. The sacred places also in which they had quartered themselves were full of corpses of persons that had died there, just as they were; for as the disaster passed all bounds, men, not knowing what was to become of them, became utterly careless of everything, whether sacred or profane.
>
> All the burial rites before in use were entirely upset, and they buried the bodies as best they could. Many from want of proper appliances, through so many of their friends having died already, had recourse to the most shameless sepultures: sometimes getting the start of those who had raised a pile, they threw their own dead body upon the stranger's pyre and ignited it; sometimes they tossed the corpse which they were carrying on top of another that was burning, and so went off.

And, eerily, Thucydides reported, not even the birds and animals would touch the corpse of a plague victim, almost as if they instinctively knew that the bodies were infectious.

Besides death, the plague brought lawlessness to a people previously known for their respect for law and decency. Looting of the dead and sick, wildness and immorality pervaded the seemingly doomed city.

Much of what Thucydides describes is suggestive of smallpox. Some of the symptoms are beyond those of variola but there is nothing that cannot be explained by the fact that Athens had no history of the disease.

Where had it come from? Whether the outbreak originated in the Middle East or North Africa or was communicated to those areas from the Indian subcontinent or the Orient is not known. What is known is that smallpox is a relatively young disease, a disease of civilization.

Because it has no animal reservoir and is not capable of long latency periods (unlike the chickenpox virus, which can lie dormant in the body for years only to emerge as a different disease, shingles), variola relies on more-or-less direct transmission between human hosts. That requires something of a population base. Without a sufficient number of humans to transmit the disease it could not persist.

It is likely that variola mutated from a virus in a domesticated animal, possibly cattle. But since it needed close, relatively large populations to keep the infection chain going, smallpox could not have become endemic until men began living in cities. Even then there were probably many false starts, with the disease breaking out in one victim only to die out when he died or when the virus itself expired before it could be transmitted to another host.

Only after centuries of attempts could the mutated virus find the right circumstances to begin an endless chain. Pinpointing that time and place is impossible.

Part of the problem is semantic. The historians in the Middle Ages who are responsible for transmitting most of the ancient texts frequently used the word "plague" to describe all epidemics because bubonic plague played such a major role in the mortality rate. But not all "plagues" were plague.

Biblical references to "plagues" represent only a later translation of terms of unknown meaning. The ancient writers could have been talking about smallpox, measles, typhoid fever or symptoms like dysentery. All these diseases were known before 500 B.C., although none had become massive and constant killers so far as can be determined.

The ancients, by and large, may have lived in some kind of balance with smallpox, probably by confining it to a childhood affliction. If one dose of the ailment, generally present in some degree, conferred lifelong immunity, eventually it would single out children, the only sure supply of nonimmune hosts. So smallpox probably existed as a childhood disease, tragic but not devastating enough to prevent population growth and the creation of empire.

There is no doubt that the Chinese, Japanese and Indians knew about smallpox very early. It could be as much as three thousand years old in Asia and perhaps it was there that the mutation into its more virulent form took place.

The foray of Alexander the Great into Asia (around 320 B.C.) may have run into the disease, suggesting an Asian reservoir. After a mutiny in Alexander's army, the Macedonian split his forces, sending part on foot across the fierce deserts and part by sea. The marching troops apparently suffered mightily. About three quarters of them died off, and, according to Quintus Curtius, during one encampment along the Ganges "a scab attacked the bodies of the soldiers and spread by contagion." We will never know if that was smallpox. If it was, however, it probably stayed with Alexander's dead and did not return with him to Europe. Perhaps European outbreaks after Pericles' time just came west in quiet, unknown caravans.

When Lucretius (c. 98–55 B.C.) translated Thucydides he

added some descriptions to the text which give strong evidence that he too knew of the disease.

A first-century Jewish philosopher, Philo Judaeus, who wrote on the Bible, described the disease occurring in Exodus as producing pustules. Some medical authorities believe that the lesions on the faces of many eleventh and twelfth century B.C. Egyptian mummies could have resulted from smallpox. Such lesions have been found on the mummy of Ramses V, who died in 1160 B.C.

Although there is no identifiable disease recorded in the annals of Rome, there were some suspicious epidemics. The worst was in 165 A.D. It was brought to the Mediterranean by soldiers returning from a campaign in Mesopotamia. Between one quarter and one third of the population died then in this "Antonine plague."

A second major outbreak of what was apparently the same disease came in 251 and lasted until 266. The death toll in that epidemic was even higher. The fallen numbered as many as five thousand a day and rural populations were particularly hard hit.

The only source for a description of the disease is the renowned physician, Galen, who was alive during the Antonine plague. His account leaves much to be desired. He said that the disease involved an outbreak in the skin and the spitting of blood, which could signify hemorrhagic smallpox. But that is a rare form of the affliction and modern experts doubt that many Romans could have had it. Additionally, Galen had a unique concept of the nature of disease in general, and he saw only what he considered germane to his concept. Since he believed that sores were nature's way of ridding the body of rotten blood and had no particular significance in understanding the affliction, he made only passing reference to their occurrence. His other descriptions of possible smallpox are scattered and inconsistent.

Besides his limited conception, Galen was also at something of a journalistic disadvantage. Unlike Thucydides, Galen followed the custom of the times and left town as soon as the disease struck. Most of Rome's doctors went with him.

Whatever its birthplace and however it came West, the first historical incident that even those inclined to doubt the nature of the Athenian outbreak agree involved smallpox was the so-called Elephant War in the year 568 A.D. It is referred to in the Koran:

> In the name of the most merciful God. Hast thou not seen how thy Lord dealt with the riders of the elephants? Did he not make their treacherous design an occasion of drawing into error; and send against them flocks of birds, which cast down upon them stones of baked clay, and destroyed them like corn trodden down by beasts?

According to the colorful legend, Abraha, an Ethiopian Christian and viceroy of Yemen, in 568, the year Mohammad was born, built a magnificent church in Sanaa to try to attract Arab pilgrims and convert them from idolatrous worship. The Arabs then had no predominant religion, although there was a strong monotheistic influence from nearby Jewish and Christian populations. Mecca, the traditional Arab place of reverence, was under the leadership of Mohammad's grandfather, Abd-al-Muttalib.

The Arabs, seeing the holiest place in Mecca, the Kaaba, in a state of neglect and perhaps blaming the Christians, stole into Sanaa one night and desecrated the church. The furious Abraha assembled a large army of men and war elephants to attack and raze the Kaaba.

Mohammad's grandfather called for heaven's intercession. Apparently it was granted, because when Abraha tried to enter Mecca all his elephants sat down.

At this moment, according to the legend, a flock of gulls,

called the *Abahil,* came flying in from the western sea. Each carried a pea-sized stone in its bill and a stone in each talon. Fittingly, every stone had the name of its intended victim on it.

The *Abahil* bombarded the Ethiopian warriors from the skies. According to the Arabic historian al-Tabari:

> Whenever one of these stones struck, there arose an evil wound, the pustules all over. At that time the smallpox first appeared.... The stones undid them wholly. Thereafter God sent a torrent which carried them away and swept them into the sea. But Abraha and the remnant of his men fled: he himself lost one member after another....

With his limbs rotting, Abraha retreated to Ethiopia with the remnants of his army to report to his emperor. All the while the gulls circled overhead, pelting the survivors with more stones. As Abraha was brought along on his retreat "his limbs fell off piece by piece, and as often as a piece fell off, matter and blood came forth. . . .

"Sixty thousand returned not to their homes," al-Tabari wrote, "nor did the sick continue in life after their return." Abraha died as he reported the tragedy to his lord.

According to some writers one elephant was also struck with the disease.

Whatever the truth contained in the legend, all historians agree that the siege of Mecca was broken by a sudden, virulent outbreak of smallpox.

It is clear that the Arabs were aware of smallpox before the Elephant War, although they were inclined to do nothing about it, particularly after the rise of Islam. They took the attitude that pestilence was the will of Allah and public health efforts to fight the disease were sacrilege. As a result of his grandfather's experience, Mohammad warned his followers to stay away from areas where there were epidemics and advised anyone living in an infected area not to travel. But he also said that "he who dies of epidemic disease is a martyr. . . . It is a punishment that God inflicts on whom He wills, but He has granted a modicum of clemency with respect to Believers."

When the Muslims later came in contact with Christians who maintained at least some public health safeguards, such as quarantine, at first they considered these efforts amusing and, later, an affront. This attitude had the effect, incidentally, of putting them at a considerable disadvantage in those places where the Muslim and Christian worlds competed. The Muslims were susceptible to every passing epidemic and their populations gained only by conversions, not by natural population growth.

But for many centuries the Arab world was unusually tolerant of intellectual curiosity and granted great liberties to those who followed the arts and sciences. It is not surprising, therefore, that their golden age produced the first complete medical report on smallpox, the classic *Treatise on Smallpox and Measles,* written by a man known to the West as Rhazes.

He was born in the Persian city of Rai probably about 865 A.D. It is said that in his thirteenth year he visited a hospital in Baghdad and became convinced that medicine would be the sole passion of his life. He enrolled under the service of the great Jewish physician Ali Ibn Sah al-Tabari. (Most of the leading "Arab" physicians were, in fact, not Arabs, but Jews, Persians and Spaniards.)

Rhazes eventually became director of several hospitals and traveled all over the Islamic world, famed for his compassion and scholarship. He wrote knowledgeably about a number of afflictions, from hiccups to life-endangering episodes of choking.

Toward the end of his life he was blinded by cataracts but

refused surgery, explaining that he had already seen enough of the world. Having given away most of his fortune to the poor, he died around 925 in poverty.

His *Treatise,* written in 910, is notable for its detail, kindness and rationality. There is no doubt about the disease Rhazes is describing:

> The eruption of the Small-Pox is preceded by a continued fever, pain in the back, itching in the nose and terrors in sleep. These are the most peculiar symptoms of its approach, especially a pain in the back with fever; then also a pricking which the patient feels all over his body; a fullness of the face, which at times goes and comes; an inflamed color, and vehement redness in both the cheeks; a redness of both the eyes; a heaviness of the whole body; great uneasiness, the symptoms of which are stretching and yawning; a pain in the throat and chest, with a slight difficulty in breathing, and cough; a dryness of the mouth, thick spittle, and hoarseness of the voice; pain and heaviness of the head, inquietude, distress of mind, nausea, and anxiety . . . heat of the whole body, an inflamed color, and shining redness and especially an intense redness of the gums.

He was sufficiently versed to know that the disease had various forms, some more serious than others. The cause of smallpox had to do with the nature of blood, he felt:

> I say that every man, from the time of his birth till he arrives at old age, is continually tending to dryness; and for this reason the blood of children and infants is much moister than the blood of young men, and still more so than that of old men. And besides this, it is much hotter. . . .

He likened the change in blood to the transformation of grape juice into wine and supposed that as people aged the blood fermented until "the blood of old men . . . may be compared to wine which has now lost its strength and is beginning to grow vapid and sour."

> . . . this is the reason why children, especially males, rarely escape being seized with this disease, because it is impossible to prevent the blood's changing from this state into its second state. . . .

His reference to the fact that smallpox hits mostly males may be an artifact of Islamic society in which women and girls are isolated in the community. It is possible, though not certain, that he reported correctly and that females suffered less.

Like many of the thousands of physicians who followed him, Rhazes was not shy about prescribing how to treat smallpox successfully. In many ways his treatment was better than that of most later physicians, and except for the hideous practice of bleeding (which may have killed more of the sick than the disease), at least was not harmful or torturous, although it too did nothing to stem the spread of epidemics.

A strict diet was vital to Rhazes' treatment. He suggested foods to assault the fever, but had some peculiar restrictions. Melons, he warned, "especially sweet ones . . . are entirely forbidden; and if the patient happened to take any, he should drink immediately of the inspissated [thickened] juices of some acid fruits. He may be allowed soft fish and buttermilk."

How Rhazes, obviously an intelligent and observant man, came to such conclusions is hard to understand. At least in regard to the high fever of smallpox, his treatment was much better than those that followed (not until Thomas Sydenham in the early 1700s was anyone as kind to smallpox patients):

> Let the patient drink water made cold in snow to the highest degree, several times and at short intervals, so that he may be oppressed by it, and feel the coldness of it in his bowels. If, after this, he should continue to be feverish, and the heat should return, then let him drink it a second time, to the quantity of two or three pints or more, and within the space of half an hour; and if the heat should still return and the stomach be full of water, make him vomit it up, and then give him some more. . . . Cold water, when it is sipped a little at a time, provokes sweat, and assists the protusion of the superfluous humors from the surface of the body.

This treatment made his patients feel better and compares favorably to the tortures inflicted by his successors.

Rhazes had several medicines that he swore worked well. Probably they did little good and must have smelled and tasted abominably. One medicine—"very beneficial and useful"—consisted of specific quantities of roses, lentils, yellow figs, gum tragacanth, white raisins (pitted), lac [the tree resin that forms shellac], sweet fennel seed and smallage [wild celery]. It was to be boiled in two pints of water until reduced by half and administered half at a time with a small amount of saffron.

To save the eyes he strongly recommended drops of rose water several times a day. Washing the face and eyes with cold water also seemed to help, he said. In really severe cases, he recommended an eyewash of red horn poppy (which probably did ease the suffering, as it contained opium) and the juice of unripe grapes, russet (an apple), aloe (lily) and acacia, mixed with saffron. In the worst cases the same medicine was to be used minus the saffron but with iron (hematite).

For the mouth sores he recommended a gargle of acidic pomegranate juice, which must have hurt like hell.

His recommendation for a lotion to ease the scars in the eyes included the "dung of sparrows, swallows, starlings, mice and crocodiles." Before that was administered he suggested that the physician lick the eyes with his tongue. After the lotion a mild aromatic powder was to be applied to the eyes. "But be sure to look carefully and frequently into the eye," he warned, "and if it be painful and red, then omit this treatment for some days."

To avoid scars on the face he recommended a medicine that included sponge, coral, almonds and kidney beans.

Everyone should be bled, even to the point of fainting, he said.

Implied in Rhazes' theory that the cause of smallpox is related to the infantile blood is the suggestion that the disease had already become basically a childhood affliction when he wrote his *Treatise.* He implied that smallpox was endemic to all of Islam, from Persia to Spain. His travels put him in a position to know, and the evidence supports him.

A childhood or endemic disease must meet three criteria: it must confer a strong, lifelong immunity so that when child victims reach adult years they cannot suffer a return bout; it must strike the unimmunized of every age, so that first contact is as bad for adults as for children; and it must be so common that the chances of being infected as a child are great.

To fit the last requirement meant that smallpox indeed was a very common disease in Islam in the tenth century. It also meant that it had been around for quite a while to have achieved that status.

We know how many years that required in other societies. The first known appearance of smallpox in Germany was in the twelfth century. It took two hundred years for the disease to become recognized as a childhood affliction. If this period applied to all societies, the Arabs had had smallpox for at least two hundred years also, since the eighth century. As the Arabs were great travelers, it is safe to assume that if they suffered from smallpox so did other peoples.

Meanwhile, in Asia chroniclers were also reporting the affliction. In 310 A.D. a plague hit the northwestern provinces of China. In Biblical fashion it was preceded by locusts and famine, and according to accounts of the time killed 90 percent of the population. It lasted two years and was followed ten years later by still another outbreak. The disease was one involving a rash and fever.

Sometime between 291 and 361 A.D. the physician Ko Hung reportedly wrote:

> Recently there have been persons suffering epidemic sores which attack the head, face and trunk. In a short time, those sores spread all over the body. They have the appearance of hot boils containing some white matter. While some of these pustules are drying up a fresh crop appears. If not treated early the patients usually die. Those who recover are disfigured by purplish scars which do not fade until after a year. . . .
>
> The people say that in the fourth year of Yung-hui this pox spread from the west to the east and spread far into the seas. If the people boiled edible mallows, mixed them with garlic and ate the concoction with a small amount of rice to help it down, this too would effect a cure. Because the epidemic was introduced in the time of Chien-wu [date unknown], when the Chinese armies attacked the barbarians at Nan-yang, it was given the name of "Barbarian pox."

The document hints of a link between the "Barbarian" invasion and the disease, which means it could have been imported into China. Because of a dating problem, the best that can be said is that smallpox was known in China sometime between 37 A.D. and 653.

The first recorded epidemic in Japan came in 552 A.D., when a considerable number of Buddhist monks from Korea came to the Japanese court. Another epidemic occurred in 585. Even worse epidemics took place in 698, 735–737 and 763–764. In 790 an epidemic struck "all males and females under the age of thirty." By 1243 smallpox had settled down to the status of an endemic disease in Japan.

If smallpox first arose in the Orient, it spread to the Middle East and then extended itself westward.

Certainly it did not stop in the Middle East. In 570 Marius, a bishop in what is now Switzerland, reported that "a violent malady with a relaxation of the bowels, and the variola, afflicted Italy and France."

Leaving aside the question of possible earlier epidemics already discussed, smallpox had finally come to Europe and it would be 1972 before it would leave, returning to its ancestral home.

It is hard to trace the course of the disease in Europe. For one thing, as has been noted, the word "plague" causes confusion. The epidemics could have been of bubonic plague, flu, smallpox, measles, typhus or any number of ailments, but the writers, none of whom were medical journalists, merely stated that in such-and-such a year many people died of a "plague."

What seem to be among the earliest specific European references in the Middle Ages to smallpox—under the name "bolgach"—appear in Irish manuscripts for the years 675, 679, 680 and 778. One typical reference is from the Annals of Ulster, which reports: "There was in the year 679, a grievous Leprosy which in Ireland is called Bolgach." In the ancient Irish tongue "bolgach" meant smallpox. The misleading mention of "leprosy" was occasioned by the common practice of calling every serious skin affliction by that name. "Bolgach" later became "galar-breac"; Irish chronicles are full of that word.

As well as "plague" and "leprosy," smallpox probably was also known as the "pestilence of fire," according to the nineteenth-century medical historian James Moore. So when Genulf wrote that in 923 "it was shocking to hear the groans of the sufferers, to see parts of their bodies, as if burnt, dissolving away, and to smell the intolerable fetor of the putrid flesh" from an epidemic of the "pestilence of fire," he was probably reporting on a smallpox epidemic.

In a chronicle of 994 we read: "The pestilence of fire burnt in Limousin, where innumberable bodies of men and women were consumed by invisible fire; and 40,000 people were killed by it in Aquitaine." And according to a tenth-century report, "The people died miserably from the limbs being burnt black by a sacred fire."

Annals are full of case histories, usually of the mighty and famous. Thus, "About Christmas, 961, Baldwin, the son of Arnolph, Earl of Flanders, was attacked with a disease, which physicians call Variola, or the Pock, and died on the day of Our Lord's circumcision following [January 1]." And a British princess, Elfreda, died of the disease in 917.

Although smallpox was not yet the dominant killer it would become—bubonic plague was by far the greatest cause of death—it was serious enough to warrant pleas to the Almighty for salvation. An Anglo-Saxon manuscript of the tenth century hailed the efficacy of this prayer:

> In the name of the Father, of the Son, and of the Holy Ghost, amen. May our Savior help us. O Lord of Heaven . . . hear the prayers of thy manservants and of

> thy maidservants. O Lord, Jesus Christ, I beseech thousands of angels that they may save and defend me from the fire and the power of the Small Pox; and protect me from the danger of death: O Christ Jesus, incline your ears to us. &C.

The manuscript provides about a dozen places where the supplicant was to genuflect.

Scarring from the disease was already commonplace and popped up frequently in descriptions of people as early as the twelfth century. One chronicler painted a word picture of a certain Count Joscelin, who fought in the Crusades and died in 1132:

> This count was small in stature, his limbs were finely formed, his hair brown, and his countenance pleasing, though pitted with marks of the smallpox; his eyes were large, and his nose aquiline. He was gallant and fierce in battle, but loved the pleasures of the table and was too luxurious.

The luxurious knight was probably also dangerous. There is strong evidence that warriors returning from the Crusades were instrumental in transmitting the disease from the Middle East to places still unaffected and spreading it throughout Europe in an age when few others traveled and contagion from human-host diseases needed assistance.

But spread it did. Danish authorities reported severe smallpox epidemics in Iceland in 1257 and 1291. A chronicle in the British Museum reports that in 1366 there "fell a sickness that men called 'ye pokkes' [which] slogh both men and women throrgh ther Enfectyne." The chronicle implies that by 1366 there was no question of smallpox being infectious.

There were outbreaks of "galar-breac" in Ireland in 1237 and 1368.

The disease mainly assumed its modern name by the end of the fifteenth century to distinguish it from the ailment of syphilis, which, by one theory, was given to Europe by the American Indians (who received smallpox in return). The first recorded outbreak of syphilis was in 1495, when soldiers returning from the siege of Naples spread the disease all over Western Europe. The English called it the "French disease" in honor of the French king, Charles VIII. (To the French it was the "Neapolitan disease.") A more acute form of syphilis than we know now, in its later stages it produced huge pustules called pocks. To distinguish variola, the designation "small" was regularly added to "pox" (hence "smallpox"), while syphilis frequently was called "great pox."

It was not until the sixteenth century that physicians could tell the difference between smallpox and measles. By that time both were already childhood diseases common all over Europe. In Alkmaar, the Netherlands, smallpox struck in 1551, and in Delft in 1562–63, hitting almost every child.

Sometime in the late seventeenth century Father Kircher, a Roman scientist, made the first microscopic examination of matter from a smallpox pustule. He believed, though he was incorrect, that he had found a *contagium animatum,* a living contagious element, responsible for the disease.

Dissertations on the cause of smallpox took up much of the time of medical philosophers. Marc Duncan, a French physician, declared in 1635 that smallpox was caused by the fear of smallpox. Gunther Schelhammer of Kiel in 1697 expressed the belief that the fear alone was fatal. Friedrich Hoffmann the Younger in 1689 said that the disease was the product of sharp alkaline salts in the system, a favorite causative agent of that century. Hieronymus Mercurialis was sure that the disease was hereditary and originated in the stars.

None of these moderns was far out of step with some of his ancient cousins. In 980 an Arab physician, Hali Abbas, wrote that the cause of the disease came from the mother during pregnancy. The fetus draws blood and milk from the mother, he wrote, taking only the best part of each and rejecting the rest. The rejected matter appears on the skin as pustules or marks. (He did note that being around a victim with festering sores seemed to bring on the same complaint.)

A later colleague, Avicenna (980–1037), remarked the obvious contagiousness. To effect a cure, Avicenna recommended opening the sores on the seventh day with a golden needle, cautioned against drafts and encouraged keeping the patient warm—which was no problem for the fevered victims.

The debate on treatment and cause ran unabated through Europe as quickly as did smallpox. In 1670, one of the great medical minds of France, Gui Patin, used the affliction as a cornerstone for his attack on Arab medicine and on anyone who tried to treat disease rationally.

> There are those who say that we must find a specific [cure]; but that is how charlatans and chemists talk, who boast of having specifics against epilepsy, quartan fever, smallpox, leprosy, gout, etc. When I hear these yarns which are worse than Aesop's Fables, it seems to me I see a man who wants to show me how to square the circle, the philosopher's stone, Plato's Republic or first matter. . . .

Bleeding, that was the treatment, Patin swore. "There are no remedies in the world that work so many miracles as bleeding." He bled all his patients during the smallpox epidemic of 1647 in Paris, and insisted he had cured them.

A German physician, Michael Etmuller (1646-1683) suggested that smallpox was *not* caused by excessive sex. As a cure he recommended a general potion of horse dung but thought sheep dung worked even better. He recommended goat manure for measles. Etmuller, one of those who crossed the indistinct boundary between science and magic, also suggested that drawing a line around the eyes with sapphires would save the eyesight of smallpox victims.

John Colbatch expounded at length in 1696 his belief that smallpox was caused by alkaline which broke up "globules" in the blood and produced purple spots on the skin, a sure sign of death in hemorraghic smallpox victims. The alkalines impeded the movement of animal spirits through the nerves. He tested treatments based on his hypothesis on five hundred patients and reported great success. If his patients had a great deal of viscid matter in the stomach, he applied a gentle emetic and then a poppy-seed or another opiate. To check the blockage in the blood and strengthen its texture, he gave a "julep" of sugar, barley water, water of cinnamon and orange and lemon juice. If the patient was delirious (too much impediment to the animal spirits), he recommended bleeding and strong acids like spirit of vitriol, niter, sulfur and barley water, along with cinnamon and sugar or syrup for flavoring. For the depressed patient he recommended cordials, which may have helped somewhat. After the sores disappear, he said that the patient should receive a gentle purgative five or six times a day. None of this, of course, did the slightest good in stemming the disease.

Matte Faveur, a chemist at the University of Montpellier, wrote in 1671 that using a combination of gold and mineral water worked wonders for smallpox victims and stopped vomiting.

The Spanish physician Forestus (b. 1522) was one of the few who thought leaving well enough alone was the best cure. He was particularly emphatic in suggesting that the physician not pick at the sores. He observed, correctly, that "when the pustules matured, and dropt off spontaneously, they were more easily cured and left fewer pits." Unfortunately he did not always practice what he preached and was found playing with the sores of some of his patients, rubbing in oil of almonds to ease the suffering.

Becoming involved in an early seventeenth century controversy that would only be settled in England many years later, Daniel Sennert was adamant in insisting that smallpox patients be kept warm, a recommendation that must have meant hell to thousands.

> Every attention is to be paid, especially in winter, to hinder the admission of cold air. The patient is therefore to be tended in a warm chamber and carefully covered up; lest by closing the pores of the skin, the efforts of nature should be impeded, the humors should be repelled, and the matter which ought to be driven out should be retained; from which anxiety, fever, and all the other symptoms would be augmented, to the imminent danger of the patient.

He did suggest, almost as an afterthought, that too much heat would "torture the patient." As we shall see, he was right.

One peculiar form of treatment involved the color red. At some time or another the idea that the color was an impediment to the spread of smallpox and an actual benefit to patients became the common wisdom. The notion possibly originated in Portugal. Literature is full of instances in which patients were wrapped in red, the bedclothes dyed red and the physician and his assistants all dressed in red.

In the late thirteenth century Gilbert, author of the first English medical text, reported that "the old women in the country added to the drink of the sick some burnt purple (or red ingredients), which, like cloth dyed in grain, had a virtue of curing smallpox."

Bernard de Gordonio in the fourteenth century recommended wrapping the patient in red. In 1335 John of Gaddensden, who treated the royal family for smallpox, wrote:

> When the son and the renowned king of England [Edward II and John, brother of Edward II] lay sick of the smallpox, I took care that every thing around the bed should be of the red colour; which succeeded so completely that the prince was restored to perfect health, without a vestige of a pustule remaining.

He also recommended feeding the patient pomegranates and other red fruits. He had all the physicians and attendants dress in scarlet.

Franciscus de Pedemontium, writing in 1300, advised:

> . . . excite and assist nature in drawing them [pustules] to the skin; [this] is to be done by warm air and by red bed coverings. . . . The blood should be carried to the surface of the body, by looking upon red substances. . . .

It probably did no worse than many of the other cures physicians suggested.

And these cures for smallpox were even embraced by the rational. Isaac Newton recommended a patent medicine called Leucatello's Balsam as a wondrous cure-all for smallpox, poisoning, rabies, wind, colic and bruises. "For the measell, plague or small pox, a half-ounce in a little broth," was recommended by the great scientist. "Take it warm and sweat after it," he added.

Leucatello's Balsam consisted of turpentine.

III

Of All the Ministers of Death

In England smallpox seemed to go through a dramatic, if subtle, change early in the seventeenth century.

Until then it was just another of the many diseases that made life an obstacle course between mortal perils. It had become one of several childhood diseases. Once a person survived childhood his life expectancy was not very much shorter than it is today, modern medical propaganda to the contrary. But some time in the early part of the seventeenth century the disease seemed to increase in frequency and virulence, thereby influencing society more profoundly. English historians, some of whom seem to have made the history of smallpox a passion, believe that they can trace this surge in the documents of the times even though the first real death records were not kept until the third decade of that century.

It is possible that the disease virus altered and became

stronger. It is more likely, however, that there was a fresh importation of stronger variola major from the Continent. Smallpox became one of life's great terrors, second only to the bubonic plague in its destructiveness. It is also possible all of this is the result of better record keeping and the disease never actually altered though it seemed to.

It was in England that the first humane treatment for smallpox victims was developed. Lives were saved—not from the disease, but from the ministrations of the medical profession. It was still the age when calling a doctor frequently did more harm than doing nothing at all.

England was also the place where the weapon that ultimately defeated smallpox was refined. But that was not until the end of the eighteenth century.

Early records in England are fairly scarce. It can be said with some certainty that the English probably knew about smallpox as early as the seventh century. We can assume that if the Swiss and the Irish had it by then, the English could not have escaped.

There was an epidemic of some kind in England in 664 A.D., but that could have been smallpox, measles or the flu.

There are very few other references specific enough to detect the presence of the disease until the *Chronicle of Brute,* which reported in 1366 that "there fell also such a pestalence that never none such was seen in no man's tyme or lyf, for many men, as they were gone to bede hole and in gude poynte sodanly thei died." The *Chronicle* blamed "ye pokkes" on an infection.

In the fourteenth-century *Piers Plowman,* William Langland noted:

> Kynde [nature] came after with many keen sores,
> As pokkes and pestilences, and such people shent;
> So kynde through corruptions killed full many.

One of the earliest sixteenth-century references is the court note that on July 14, 1518, Henry VIII left Wallingford because of the smallpox, measles and plague simmering there.

In Bullein's *Sickness and Health,* of 1562, there is a mention of a "medicen" for smallpox and a lotion "to annoint the faces of children that have the small pockes, when the said pockes be ripe, to keep them from pittes or erres." Later in the century Willis wrote "that this disease creeps on by contagion [which] is shown by daily experience."

Mention of the disease had begun to creep into the diaries of the time and it becomes possible to begin piecing together some of the personal stories of its victims from the mid-sixteenth century. One of the earliest is the "deathbed confession" of one Richard Allington, Esq., who was something of a financial scoundrel. On November 22, 1561, he found himself, to his surprise, dying of smallpox. The ailing Allington summoned the Master of the Rolls and four lawyers to his bed and said:

> Maisters, seinge that I muste nedes die, which I assure you I nevar thought wolde leave come to passe by this dissease, consydering it is but the small pox, I woulde therefore moste hertely desyre you in the reuerence of God and for Christes passions sake to suffer me to speake untyll I be dede, that I may depcharge my conscens.

He then confessed that he was a usurer and admitted having too little appreciation of God's graces. He urged the lawyers to make sure that all those with whom he had dealt were taken care of, particularly a hundred who repaid him at unusually usurious interest rates. Satisfied then that he had cleared his conscience, he asked that the Psalms be read.

A witness wrote: "And then he thought he should have died, but then broth being given into him, he revived again and fell to prayer and gave himself wholly to quietness."

Allington did die a few weeks later, but of something else. His hundred victims were never paid back and his widow reportedly lived on in considerable wealth.

Elizabeth I survived a bout with the disease in 1562. The episode left her with a healthy respect for smallpox, and when it broke out near court or near the residence of someone she cared for, she would urge flight.

By 1593 smallpox was being described in detail in English medical books such as Simon Kellway's, and sometime in the next fifty years apparently it increased in its effect, becoming the major killer it would remain for several centuries.

In 1602 the poet Thomas Spilman was moved to write this ode to the disease:

UPON HIS LADIES SICKNESSE OF THE SMALL POX

. . . Are not these thy steps I trace
In the pure snow of her face?
Th' heavenly honey thou dost sack
From her rose cheeks, might suffice;
Why then didst thou mar and pluck
Those dear flowers of rarest price?

Ben Jonson wrote: "Envious and foul disease, could there not be one beauty in an age, and free from thee?"

George Bell of Edinburgh reported: "The smallpox . . . ever since its introduction into Europe, more than a thousand years ago, has descended with undeminished violence from generation to generation. . . ."

Correspondence of the times reflected the growing horror and the diverse ways it was handled. The great poet John

Donne, dean of St. Paul's cathedral in London, wrote in 1631:

> At my return from Kent to my gate, I found Peg had the pox; so I withdrew to Prickham and spent a fortnight there. And without coming home, when I could with some justice hope that it would spread no further amongst them (as I humbly thank God it hath not, nor much disfigured her than had it) I went into Bedfordshire. . . .

On the other hand one anonymous letter of 1624 says that ". . . the Lady Purbeck is sick of the smallpox, and her husband is so kind that he stirs not from her bed's feet."

A poignant letter came from the house of Sir John Coke of Garlick Hill, London, written by his wife:

> It pleased God to visit Mrs. Ellweys [a stepdaughter] with such a disease that neither she nor any other of her nearest and dearest friends durst come near her, unless they would hazard their own health. The children and almost all our family were sent to Tottenham before she fell sick, and blessed be God are all in health. Mrs. Ellweys was sick with us of the small pox, twelve days or thereabouts.

Mrs. Ellweys was also pregnant, and before the disease abated she went into heavy labor. She died on June 15 at about 5 a.m., apparently after a stillbirth. She was buried hurriedly at 10 that night with only two members of her family at the grave.

> God knows we have been sequestered from many of our friends' company, who came not near us for fear of infection, and indeed we were very circumspect, careful,

> and unwilling that any should come to us to impair their health.

The saddened Lady Coke lamented that during the episode she could not even visit her children for fear of infecting them.

Royalty was hardly immune from tragedy. When Charles II left The Hague on May 23, 1660, to assume the throne of England, he took two of his brothers with him, the Duke of York and the Duke of Gloucester. In the early autumn Gloucester fell ill at Whitehall. A few days later, the smallpox having come out "fully and kindly," he was vastly improved. At about 6 p.m. the royal physicians, thinking there was nothing much they could do for him, left the bedchamber.

Shortly after their departure Gloucester began bleeding copiously from the nose, fell into a deep coma and died. According to that veteran gossip, Samuel Pepys, the physicians were widely blamed for doing nothing to save him. An autopsy revealed that Gloucester's lungs were full of blood, "besides three or four pints that lay about them, and [there was] much blood in his head, which took away the senses."

Three days after the royal funeral the King and York went into Margate to greet their sister, the Princess Mary of Orange, on her arrival from The Hague. About the middle of December she too was stricken with smallpox and died on December 21. Again Pepys reports that the doctors were widely blamed. ". . . now they [people] strike not to say, with your Agrippa [Pepys himself], that at least in these, a physician is more dangerous than the malady."

Pepys was right. Physicians had the choice of making their patients more comfortable or providing them with excruciating torture. Most unknowingly chose the latter. Smallpox victims were burning up with fever, but centuries of physicians warned against anything to cool their bodies.

A draft was considered more dangerous than the disease. So, despite the high, agonizing fever, the common treatment for smallpox was a closed, stifling room, piles of bed clothing and a roaring fire. Frequently guards were placed around the victim to prevent him from doing anything to feel less distress from heat, such as opening a window. Like the Athenians of Thucydides' time, the patients wanted most to jump in a rain barrel or a stream. That was the thing doctors were least likely to let them do.

Gradually, however, the heat treatment came under attack. The man who led the medical establishment toward a more humane, and probably more effective, treatment was Thomas Sydenham.

Sydenham, born in 1624, was the eighth of ten children of William Sydenham, a wealthy landowner. He enrolled in Oxford in 1642, but the country's increasing polarization between the Royalists and Puritans made study impossible. He finally joined the Puritan forces with the rest of his family and rose to the rank of captain. He received a medical degree in 1648 and was appointed to the faculty of All Souls' College. Sydenham continued there—with one more brief return to the military—until 1655, when he set up private practice, and qualified for the Royal College of Physicians in 1661.

British medicine during the Restoration was a wild free-for-all, with traditional physicians competing against armies of quacks and charlatans, many of them "chemists" with secret formulae, many invoking spiritual remedies. It was in this atmosphere that Sydenham rose to the fore of his profession.

In an era of dandies he was a simple Puritan. A large man with a reddish complexion and gray eyes, he had long hair, which he wore without the usual wig. As a practitioner he was widely respected by patients and colleagues alike. He had a very low regard for the conventional wisdom. Once, when asked by a student, Richard Blackmore, which books to read to qualify himself for medicine, Sydenham replied, "Read *Don Quixote;* it is a very good book. I read it still."

As a physician he demanded strict obedience to his orders. He was a pioneer in the use of quinine for fever and as a tonic, and helped to develop liquid opium, a more effective painkiller than the solid form usually employed.

When the bubonic plague broke out in London in 1665, Sydenham did follow the common wisdom—he fled. Although he felt a doctor's place was at the patient's bedside, he warned that physicians were "not exempt from the common lot, but [are] subject to the same laws of mortality and disease as others." He spent his exile writing a book on fever, which has become a classic. In later years he became a close friend of the philosopher John Locke, who frequently went on patients' rounds with him.

In 1676 Sydenham's *Medical Observations* launched the science of epidemiology, and he published his work on smallpox in 1682. It was the best treatise since Rhazes'.

Sydenham found a direct relationship between the severity of the pustules on the face and the severity of the disease. He also noted that the harder physicians worked to save a life, the less likely they were to succeed. The poor, neglected and untreated generally survived at a higher rate than the wealthy who had physicians crawling around their bedposts. This observation did not much endear Sydenham to his colleagues.

Sydenham believed that it was folly to increase the searing temperature of his patients, and he banned the hot bed. He told his smallpox patients to get up and walk around as long as they could. When the fever finally drove them to bed, he ordered the windows opened to let in fresh air. Bed coverings were to be light.

Sydenham frequently told the story of a very fat young man who was stricken with smallpox while on his way to

Bristol by stage. The fever made him delirious. He was put to bed at an inn under thick blankets and soon sank into a coma. Everyone thought he was dead. Because of the summer heat, the vile sores and the fact that he was very hard to carry, they took the "corpse" from the blanket-covered bed, put it on a table and covered it with a thin sheet. The effect was wondrous, Sydenham said. The youth came to life quickly and recovered. Sydenham concluded that the young man's life was saved by merely getting him out from under all those blankets.

Sydenham also banned heavy foods, such as meat and wine. He did allow some cool beer but generally recommended a diet of gruel and barley water. Like everyone else, he thought bleeding did some good and practiced it in moderation.

It was Sydenham's opinion that smallpox was an "epidemic constitution of the atmosphere" due to unknown natural laws, and nothing more arcane.

Actually he did not invent the cooling treatment. (It was mentioned in a play by Beaumont and Fletcher in 1626.) But he did bring it to the attention of the rest of his profession, and the number of lives saved from aggravating the fever and the torment avoided are beyond counting. He was an empiricist and demanded that facts be supported by evidence, a consideration that did not impress his peers.

On his death in 1689 he was immortalized as the father of clinical medicine, the medicine that pays less regard to philosophy and sophistry and attends to the well-being of patients. Thousands of smallpox victims owed their survival to his empiricism.

Sydenham's writings did not end any of the arguments about the disease, and the efficacy of contemporary treatment led to endless debates in the Royal College. The next treatment to come under serious scrutiny was bleeding, which proved to be something of a losing battle.

In 1718 a Dr. Woodward, well known both as a physician and a geologist, wrote an article criticizing bleeding and aiming his barbs particularly at two of his most illustrious colleagues and bleeders, the Drs. Mead and Freind. A feud began, and on June 10, 1718, at 8 p.m., as Woodward was walking through the quadrangle at Gresham College he was set upon violently by Dr. Mead.

Woodward drew his sword and challenged Mead to do the same. Mead took his time, which infuriated Woodward further. When the duel finally began Woodward, now almost out of control with rage, tripped and fell. Mead, suddenly finding himself victorious, stood over Woodward with his sword pointed at the helpless man's neck and demanded that he beg for his life. Woodward refused, whereupon the two men lapsed into a shouting match that was finally broken up by their friends. The affair was a public scandal, but did smallpox patients little good—they were bled routinely for another century or two.

Reverting to the seventeenth century, in the epidemic of 1667–68 the mistress of Charles II, the beautiful Frances Stewart, Duchess of Richmond, came down with smallpox and was "mighty full of it." Pepys had seen a portrait of her made shortly before she became ill. "It would make a man weep to see what she was then, and what she is likely to be, by people's discourses now," he lamented when he received news of her affliction. She did recover, however, and one day in August, 1667, Pepys saw her driving around the park. She was, he reported after chasing her coach, "of a noble person as ever I did see, but her face, worse than it was considerably by the smallpox." It didn't seem to matter much to the King. Pepys says that once in May while she was still recovering, he was seen climbing over the garden wall to get at the Duchess.

One of the first documented epidemics in England occurred earlier, in 1628. In the week of May 24, forty-one died of smallpox in London; the next week, thirty-eight; by the third week of June, the death toll was listed at fifty-eight. That was out of a total population of about three hundred thousand.

In 1629, the next year, the English began keeping death records, and it became possible to see just how serious a threat to life smallpox had become.

Between 1629 and 1661 an average of 596 people died of smallpox in London annually. Some years it hardly counted, in others it counted very much. The overall population of the city remained fairly constant during this period.

YEAR	SMALLPOX DEATHS	TOTAL DEATHS	% OF DEATHS
1629	72	8,771	.8%
1630	40	10,554	.3%
1631	58	8,532	.6%
1632	531	9,535	5.0% *
1633	72	8,393	.8%
1634	1,354	10,400	13.0% *
1635	293	10,651	2.7%
1636	127	23,359	.5%

* Epidemics

Notice that in some years, like 1629 and 1630, smallpox barely made a dent in the death rolls. But in others, particularly 1634, an epidemic apparently raged through London and 13 percent of all the deaths recorded in the city were attributed to smallpox. The erratic timing of the outbreaks continued through the century.

The great epidemic of 1641 is lost on the registers, but on August 26 of that year, a correspondent wrote: "Both Houses [of Parliament] grow very thin by reason of the small pox and plague that is in town, 133 dying here this week of the plague and 118 of the small pox, 610 in the whole of all diseases." (Twenty percent of all the deaths were caused by smallpox that week, second only to bubonic plague.) On September 9 a letter from Charing Cross said: "Died this week of the plague, 185, and of the smallpox, 101."

The rolls resumed in 1647, and the smallpox death rate bounced up and down.

YEAR	DEATHS	YEAR	DEATHS	YEAR	DEATHS	YEAR	DEATHS
1647	139	1648	401	1649	1,190 *	1650	184
1651	525	1652	1,279 *	1653	139	1654	832
1655	1,294 *	1656	832	1657	835	1658	409
1659	1,523 *	1660	354	1661	1,246 *	1662	768
1663	411	1664	1,233 *	1665	655	1666	38

* Epidemics

In the twenty years between 1661 and 1680, 5.6 percent of all the deaths recorded in London were due to smallpox, as near as physicians could tell. In the next twenty years, the average was up to 6 percent. During the forty years before 1700, an average of 1,257 persons annually died in agony with the disease.

In a few of those years it was almost rendered insignificant in the city by the plague. In 1665, when the plague was at its historic height, 655 people died of smallpox while the overall death toll in the city was a monstrous 97,306! By 1681 more than a thousand were dying of smallpox in London each year. If the disease was variola major, which appears likely, there was a 40 percent fatality rate, which means that several thousand came down with the disease every year, with all the suffering that entailed.

The worst years were in 1674, when, the records claim, confluent smallpox raged, and in 1681 when it returned. In many cases the victims had been weakened by earlier diseases. Creighton, in his *History of Epidemics in Britain,* said that in 1674 an epidemic of measles struck in the first six months and was the main cause of the high smallpox death toll in the second half of the year.

The year 1681 followed two very hot summers and an epidemic of infant diarrhea. That summer was just as hot, and deaths by smallpox in the last week of August reached 168.

In 1685 smallpox was uniformly distributed throughout the year and it went hand-in-hand with a typhus epidemic. The week ending September 29, 114 people died of typhus, the highest for the year. During the week ending August 18, 99 died of smallpox, the worst week of that year.

There is ample evidence that these figures for London were matched by similar outbreaks in the rest of England. While the disease was hitting its century high in 1681 in the capital city, places like Norwich and Halifax also were reporting epidemics.

Creighton reports on one man named Thoresby, whose friend lost three children in Halifax in 1681. In 1689 Thoresby himself lost both of his children in Leeds within a few days. In 1699 he lost two of the four children born to him since the last outbreak.

In 1674 the academic year was canceled at Cambridge because of smallpox.

The next century was no better.

YEAR	SMALLPOX DEATHS	TOTAL DEATHS	% OF DEATHS
1701	1,099	20,471	5.4 *
1702	311	19,481	1.6
1703	398	20,720	1.9
1704	1,501	22,684	6.6 *
1705	1,095	22,097	5.0
1706	721	19,847	3.6
1707	1,078	21,600	5.0
1708	1,687	21,291	7.9 *
1709	1,024	21,800	4.7
1710	3,138	24,620	12.7 *

* Epidemics

In the next decade 6.5 percent of all the deaths in London were caused by smallpox, with major epidemics in 1714 and 1719.

The early eighteenth-century English outbreaks had a fascinating result: a great number of people began to blame the medical profession for the death rate. One chronicler reported that in 1714 the general cry was that the cause of the high death rate during an epidemic was "the great want of help, care or advice therein." Woodward (of dueling fame) blamed bad medical treatment. Another writer said that the epidemic of 1710

> . . . was not due to a peculiar state of the air, but to a defect in some of our great physicians who being too fully employed, could not give due attendance to all or even to any of their patients through the multiplicity of them: for want of which, and the severity of their injunctions, which hindered others from applying anything in their absence, many persons were lost who might otherwise have been saved with due care.

The writers presumed doctors could really fight—and vanquish—the disease. The error would actually appear to be in their attempts to help. Those who went without physicians may very well have been better off for it.

One clue, as Sydenham had reported earlier, was the fact that the higher classes, those who presumably could afford medical care, seem to have suffered more than the untended poor.

The rich were, in fact, the prime victims of variola. In 1720 the Duchess of Argyll wrote to the Countess of Bute to congratulate her on the birth of her third child, and she added the comment: "He that has had the smallpox [is] as good as two [children], so mortal as that distemper has been this year in town was ever known."

The servants of the rich seemed to provide a conduit to the lower classes for the virus. Eventually the rich got together and built England's first smallpox hospital (1746), just for their servants. Class distinctions would rise again when the vaccination debate erupted at the end of the century.

What was true of England was true of cities on the Continent, which were also beginning to keep records of deaths. In Geneva the records were broken down by age, which gives us a full picture of the nature of the disease and how it destroyed children most of all. The records are for the years 1580–1760.

AGES	TOTAL DEATHS	% OF TOTAL
0–1	6,792	26.8
1–2	5,416	21.4
2–3	4,116	16.2
3–4	2,826	11.1
4–5	1,928	7.6
5–6	1,325	5.2
6–7	944	3.7
7–8	543	2.5
8–9	454	1.8
9–10	354	1.4
10 +	560	2.2

In Copenhagen, from the years 1750–1800, 12,309 people died of smallpox out of a total number of 173,080 deaths. That was 14 percent of the population, or one death out of every seven attributable to variola.

In Sweden from 1774 to 1798 there was a total of 4,131 smallpox deaths, four fifths of them occurring among children under five, nine tenths among those under ten. Of all the deaths recorded in Sweden at that time 7.8 percent were caused by smallpox.

One city that seems to have suffered extraordinary amounts of torture from the disease was Glasgow.

YEAR	SMALLPOX DEATHS	TOTAL DEATHS	% OF DEATHS
1783	155	1,414	11
1784	425	1,623	26.2
1785	218	1,552	14
1786	348	1,622	21.5
1787	410	1,802	22.7
1788	399	1,982	20.1
1789	366	1,753	20.9

That phenomenal death rate continued into the next decade. Of the 17,475 deaths in Glasgow during that period, 3,302, or 18.9 percent, were attributed to smallpox. This incidence of almost one out of every five deaths is the highest known rate in the eighteenth century in Europe. Liverpool came close during the three-year period 1772–74, with 18.2 percent.

In Berlin the death rate was 8.3 percent between 1758 and 1774 and 7.8 percent between 1783 and 1800. In Leipzig it was 4.2 percent for the same periods. In one other German city there were 817 deaths reported in a single year, but only one case involved a person over the age of fifteen.

These figures show only the attacks that were fatal. They do not include those pitted and blinded by variola. Historian Edward J. Edwardes reported that "the disease was regarded as universal, or almost universal." Most children in London had the disease before they were seven. Edwardes wrote:

> In Chester, in 1775, after an epidemic of small-pox, of 14,713 inhabitants only 1,060 [7 percent] had *not* had small-pox. Here the cases were 1,385 with 202 deaths

> (180 under 5 years) *all under 10* years of age. In Hastings (1730–31) there were 705 attacks in a population of 1,636; of the remaining 931 no less than 725 were survivors from previous attacks. Out of 1,250 cases of small-pox in three Prussian towns in the year 1796, the age-class under 10 years furnished 1,184, i.e. 94.5 percent. The collective population was 13,329, and out of the 12,079 not attacked in 1796, all but 524 were survivors from previous epidemics.
>
> About one-sixth or one-eighth of the attacks were fatal.

One in seven stricken with smallpox in Hastings in 1730–31 died. One in six died in Chester in 1775. In London's smallpox hospital in 1777, "a fourth died . . . in 1796 a third . . . in 1781 two-fifths. . . ." In the same city it was noted that one out of every fourteen children born eventually died of smallpox.

Perhaps the most grisly record belongs to the town of Oldenburg, Germany, 1795, where 550 out of the town's 600 children were stricken and 144 (26 percent) died.

One historian estimated that 400,000 people died of smallpox every year in Europe and that 80 percent of the populace at one time or another caught the disease.

Sooner or later it may have hit everyone. It certainly did not ignore the famous and the powerful.

IV
"Loving Her As I Do"

Elizabeth I

On October 10, 1562, the twenty-nine-year-old Queen of England was at Hampton Court working on a letter to Mary Stuart designed to reconcile their differences. Later in the day Elizabeth felt ill and, following a fad of the times, she decided to take a bath. Generally bathing was done out of a basin, using washcloths, but the upper classes had recently adopted whole-body immersion as a form of pleasure and a tonic. Elizabeth was very fond of the practice and hoped it would make her feel better. She bathed fully and afterwards went for a walk in the garden. It was a brisk day, and the Queen caught a chill and a fever.

Her doctors ordered her to bed and, to everyone's surprise, she followed their orders. She worked on what state papers she could, including the letter to Mary, but the fever mounted and it became more and more difficult to proceed.

Her cousin, Lord Hunsdon, recommended his physician,

Burcot. But when the German doctor diagnosed smallpox, she flew into a rage, shrieking, "Have the knave away out of my sight!" Burcot stormed from the room.

On October 15, despite the mounting fever and increasing pain, she tried to finish the letter to Mary.

She wrote that the English invasion of France to support the Huguenots was designed to "guard our houses from spoil when our neighbours' are burning." She said she knew that the boy King of France was "only a King in title":

> I cannot suffer such evils, as a good neighbour. You shall have no occasion to charge me with deceit having never promised what I will not perform. If I send my people to these foreign ports, I have no other end than to help the King. . . . My hot fever prevents me writing more.

That was all she could do, and she finally ceased fighting the raging fever and her terror. She would lie in bed and, periodically, at great effort raise her hands to her face so that she could see that none of the telltale marks had appeared on her clear skin. The effort was so painful that she lay panting for long minutes afterwards. She lapsed into near-comas, her breathing so labored that the guards outside her room could hear her.

The crisis in France was now replaced by a closer, far more serious one. Elizabeth was going to die without an heir, but with a cousin to the north who would like only too well to become Queen of both Scotland and England. Mary had Catholic allies all over the Continent prepared to help her.

The Privy Council was called to her bedside. The council waited patiently while the Queen slept a terrible, deep sleep. After several hours she regained consciousness and gave instructions to her councilors.

They sympathetically soothed her, promising to follow her instructions implicitly. But if she died now, they knew, England would be thrown into a religious civil war and the English Reformation of Elizabeth's father, Henry VIII, would be in mortal peril. Hunsdon was determined to save his cousin at all costs, and sent two servants to fetch Burcot.

The physician could not be forced to return until one of the servants threatened to cut his heart out on the spot if he did not.

"Almost too late, my liege," Burcot said as he entered the Queen's bedchambers and saw her shriveled, sweating body. He ordered a mattress to be placed in front of the fire and, wrapping the Queen like a mummy in scarlet flannel with only her left hand uncovered, had her lie near the fire. He gave her a drink from a small bottle. Elizabeth whispered that the potion was "very comfortable."

Soon the heat seemed to draw red spots out on her exposed hand.

"What is this, Master Doctor?" the frightened woman asked.

"'Tis the pox," Burcot replied. Elizabeth began to moan, but Burcot cut her short.

"God's pestilence! Which is better, to have the pox in the hands, in the face, or in the heart and kill the whole body?"

The pustules formed on her body, but Elizabeth did not die. Her illness produced a different tragedy.

Her nurse, Lady Mary Sidney, soon came down with smallpox, and the disease left her face horribly disfigured. Even her husband was appalled.

"I left her a full fair lady, in mine eyes at least, the fairest," Sir Henry Sidney said, "and when I returned I found her as foul a lady as the smallpox could make her, which she did take by continued attendance on Her Majesty's most precious person."

The price was bitter. Lady Mary spent the rest of her life

in darkened rooms in the Sidneys' home. She never again appeared in public, and Elizabeth never saw her again.

It took six weeks for Elizabeth to recover from the disease. One of her first actions was to reward Burcot by giving him lands and spurs once owned by her father.

The cost of the disease to the Queen is still debated among historians. It is generally believed that the high fever caused most of her hair to fall out and she was never seen without a wig for the rest of her life. She was probably pockmarked, but because Elizabethan women wore heavy, almost masklike makeup it is not known for sure how badly Elizabeth's face was marred. (Once, when she was an old woman, her current favorite, Lord Essex, barged into her bedchamber at Nonesuch without knocking, catching the Queen without her makeup and wig. He almost lost his head for his impetuosity.) Meanwhile, the French expedition was becoming a disaster. The Catholic French Queen Mother, Catherine de Médicis, called a conference of the warring French parties and, without consulting Elizabeth, ended the bloodshed. She ordered the English out of Newhaven (Le Havre), but instead they sent reinforcements, who found the Newhaven garrison dying of bubonic plague. The English all finally surrendered and returned home, carrying the plague with them. Twenty thousand died. Peace was declared the next July.

Mary II

There were several omens, and the fall of 1694 was not a happy one for Queen Mary II, who ruled England jointly with her cousin and husband, William, Prince of Orange, the reign of William and Mary.

William had returned from a campaign on the Continent, and Mary had gone to meet him at Rochester. She was suffering from a cold and he was quite sick with one himself. The two received a tumultuous reception in London and presumed that all they needed to restore themselves to good health was a rest.

A month later Mary was ill again with another cold. As was her custom she drank a medicine called Venice treacle, which caused sweating and broke her high fever. It seemed to work.

Mary went to Hampton not long before Christmas to order furniture for the new building there. William had caught still another cold and was working very hard. He did not join her, staying instead at Whitehall.

Mary, afraid that her husband would have to return to the Continent, fell into a depression. Her friends were frightened by her sadness, but she brushed off their concern. On December 19 she was not feeling well, but did not tell anyone and kept to her business. By nightfall she was worse. She went to her bedroom and took more of the Venice treacle, in hopes of breaking the fever. Motivated by some new premonition, she began putting her papers in order, burning many of them, including her diary of that year. She wrote out instructions that her funeral was to be simple, with "no extraordinary expense," and requested that no autopsy be performed. She also left written instructions for repayment of several small debts and for a bequest to two children she had befriended—haircuts for the boy, laces for the girl. (She and William were childless.)

In a letter that she left for William locked in a desk by her bed there was a "strong but decent admonition to the King, for some irregularity in his conduct." It was probably a reproach for his affair with Elizabeth Villiers, the plain but witty friend of the family whom William had used as a sort of spy in court and as a lover. One night Mary had sneaked through the halls and caught her husband leaving Elizabeth's bedroom at 2 a.m. He was angry for being spied upon and the two had quarreled bitterly. His infidelity had added to the burden this once gay young woman bore, bending her further toward melancholy.

The morning of the 20th Mary found that the fever had not eased, so she took a double dose of the treacle. It too failed to bring on the perspiration or break the fever. William, who had been informed and was now greatly alarmed, called in the doctors. One, Dr. Walter Harris, a follower of Sydenham, believed that her reliance on the treacle may have been a fatal mistake.

Smallpox was spreading throughout the area and both William and Mary were terrified. William had lost his mother to "the inexorable and pitiless distemper" fourteen years before, almost to the day. His father had died of the disease. Mary came from a family that seemed similarly prone to smallpox, which had killed her aunt and her uncle, the Duke of Gloucester, at Whitehall in 1660. The Duke had succumbed to the dreaded hemorrhagic variety. Besides, she and all of the other children of James II were rather sickly, suffering from what one physician called *mala stamina vitae.*

William had to leave her side to attend to business. He was so distracted that he could not veto legislation in Parliament which he had long opposed. During the debate he took aside Gilbert Burnet, the Bishop of Salisbury, and burst into tears. He told Burnet that the Queen probably had smallpox. Burnet wrote:

> He cried out that there was no hope, and that from being the happiest he was now going to be the miserablest creature on earth. He said that during the whole course of their marriage he had never known one

single fault in her; there was a worth in her which nobody knew besides himself.

If William knew that it was smallpox, the royal physicians were not sure. Diagnoses varied from measles to scarlet fever to spotted fever to smallpox. The problem was that Mary's symptoms were peculiar. It was not until after a prolonged period that a rash accompanied by a harsh cough appeared, but the rash did not readily fit any known pattern. The next day the doctors finally saw smallpox "under its proper and distinct form." But later the rash changed and they were thrown back into confusion.

Meanwhile William moved a bed into the Queen's room so that he could be beside her. News of her illness was spreading throughout the court and London.

"Never was such a face of universal sorrow seen in a court or in a town as at this time," Burnet wrote. "All people, men and women, young and old, could scarce refrain from tears." They crowded anterooms waiting for word from the bedchamber. They were exultant when, on Christmas Day, the rash began to disappear and Mary went into remission. The doctors, following the practice of the time, continued to bleed the poor woman and to apply leeches.

The fading of the rash seemed a good sign to them, but by Christmas night it should have been clear that the worst had happened. The disease was not going to erupt into the usual pustules, but the virus was working deep inside her.

The day after Christmas large red spots, like measles, appeared on her breast. By nightfall they had spread to her face. The next morning a physician pierced one of the marks and blood erupted. By then all doubts were settled. Mary had hemorrhagic smallpox and, in the words of Dr. Harris, she was "a dead woman."

William rarely left her side. He paced about the bed,

furious and distraught. In a letter he took time to write to a friend, he explained his feelings:

> You can imagine the state I am in, loving her as I do. You know what it is to have a good wife. If I were so unhappy as to lose mine, I should have to withdraw from the world, and though we have no monasteries in our religion, one can always find somewhere to go and spend the rest of one's days in prayer to God.

"The King's affection was greater than those who knew him thought his temper capable of," Burnet wrote. "He went beyond all bounds in it."

Mary sank in and out of comas. Once she awoke and found William weeping.

"Why are you crying? I am not so bad," she assured him.

She was in no pain but she was having great difficulty breathing and she had begun spitting blood. Blood also began to appear in her urine.

One huge red blotch appeared over her heart. Harris said it looked as if she had been burned by a hot coal. It, too, was full of blood.

Archbishop Tenison comforted the royal couple but finally he urged William to tell his wife how dire her condition was. He began to break the news, but by this time Mary sensed what was happening and finished his sentence for him.

"She thanked God she had always carried this in her mind, that nothing was to be left to the last hour," Burnet wrote. "She had nothing then to do, but to look up to God, and submit to his will; it went further indeed than submission; for she seemed to desire death rather than life."

Mary was no longer feeling very ill. She could sit up and drink broth that the doctors gave her. But she was not always rational and Burnet wrote that her mind obviously was going.

William was in agony. He fainted several times and broke into "most violent lamentations."

On December 27 Tenison gave her the sacrament with all the other bishops in the palace attending. Mary was awake during the service and repeated the words.

William wept so that Mary sometimes lapsed into prayers to avoid contact with him. He cried loudly and even told her "that if God caused this blow to fall upon him everything would be over for him." He was always considered a cool, almost cold man, and his lamentations stunned those in the room.

The doctors tried to get the disease back on the skin, to make it form the pustules that meant they had a chance to save the Queen, but it was to no avail. She began to hallucinate, remembering scenes when her sister Anne had smallpox.

Several times William had to be led from the room into an antechamber, where he collapsed.

On December 27, the eighth day of her illness, she lapsed into unconsciousness. William had to be taken from the room again. He fainted when told that her pulse was failing.

At 1 o'clock in the morning of December 28 Queen Mary II died at the age of thirty-two.

William was so distraught that it was feared that he had lost his mind. Burnet wrote: "His spirits sank so low that there was great reason to apprehend that he was following her; for some weeks after he was so little master of himself that he was not capable of minding business or of seeing company."

Because her letter was not found at the time, Mary's funeral was the elaborate ceremony that she had hoped it would not be. Her bowels were removed and buried in the chapel of King Henry VII. Her body lay in state at Whitehall.

William eventually had the buildings he and Mary had been erecting at Hampton destroyed. He never married again.

The rather juvenile verse printed below Mary's official funeral drawing told much about her subjects' love:

> With mourning pen and melting eyes,
> With bleeding heart, and sobbing,
> Here lament the love of one
> Who was the brightness of the throne.
> Our loss is her eternal gain,
> And yet we cannot complain
> At having lost the sweetest Queen
> As ever in the realm was seen

Louis XV

The reign of Louis XV, once his minority was over, had lasted fifty-one years. Only that of his great-grandfather, Louis XIV, the Sun King, had been longer.

The Bourbon crown was at its apotheosis when he had inherited it. He was a monarch by divine right, Louis the Well-Beloved, master of a country of boundless strength and energy. There were problems and crises, but nothing French kings had not been able to overcome in the past, and the young, dashing King gave the early impression that he would not falter.

Now he was a tired old man who had become Louis the Well-Hated. He had lost his empire, fought with the Parlement, and so depleted his treasury with excesses at court that the King of France could not pay his bills.

Louis had been unable to cope with a changing world. He was locked into a body of ritual and court etiquette that made most of the palaces of Europe look like peasant camps. When he was seven years old he had a hundred servants. As King it took seven men to serve him lunch.

He was frequently bored and, probably, clinically depressed. For amusement he sought women and, later, teenage girls. Even in eighteenth-century France his sexual appetite and the arrogance of some of his lovers were a scandal.

In many ways he doomed his grandson to the guillotine.

At the age of sixty-four his digestive organs did not work, his wine had to be watered, and his mind was failing him.

"I see that I am no longer young," he told his physician, "and I must begin to rein in the horses."

"Sire," the doctor replied, "you would be better advised to unharness them."

One diplomat wrote to Maria Theresa of Austria:

> The King is growing old, and from time to time seems to have regrets. He finds himself isolated without aid or consolation from his children, without zeal, attachment, or fidelity from the bizarre assemblage composing his Ministry, his society, his surroundings. . . .

It was in the depths of such a depression that Louis went to the greater privacy of one of the Trianon pavilions on Tuesday, April 26, 1774, leaving most of the court at the main palace in Versailles. He took only his current mistress, la Comtesse du Barry, his First Gentleman of the Bedchamber, the Duc d'Aumont, and a few retainers.

The next morning he felt unwell. He thought that the fresh air and the excitement of a hunt would do him good. But he was too ill to ride this time and had to follow the hunt from a carriage. At 5:30 he returned to the Trianon pavilion and went to bed.

He spent the evening with Du Barry, who administered some mild medicines, and then he fell into a fitful sleep. The next morning he was still ill and Du Barry called for a doctor. The physician, Lemonnier, found the King with a high fever, but thought it would pass. He, Du Barry, and the First Gentleman agreed that the King ought to stay at the pavilion and that word of his illness should be kept secret from the court.

Du Barry felt that the King was likely to be better off in the airy rooms of the beautiful little palace with few to bother him than in the main edifice, where he would be surrounded by the hordes of the court, a half dozen doctors, his family, and any passing Frenchman who wandered in. (Everyone properly dressed was allowed to enter the principal palace.)

The second reason was personal; Du Barry had a problem. Many years earlier, in Metz, the King had become so sick he had feared for his life. As a condition for hearing confession he had been required to send his current mistress from his home and promise never to see her again. He relented, but as soon as he recovered he called her back, touching off another scandal. Du Barry didn't want to take any chances. As long as priests were kept from Louis, there would be no repeat of Metz.

Had her plan to keep him at the Trianon pavilion worked, Louis might have survived. But the French court had too many ears, and news of his royal indisposition traveled quickly. The Dauphin dispatched La Martinière, whom Louis respected, to get the King to agree to return to the court.

He was placed in his daughter's bedroom until his own room was prepared. The King had a bad night. The fever was still up and his head felt about to burst. It was decided to call in more doctors after Lemonnier bled him.

Du Barry suggested two of her own physicians, Lorry and Bordeu. Lemonnier called in Lassonne, who had treated Mesdames. They all crowded around the sick man.

Soon the poor King's bedroom was jammed. Fourteen members of the Faculty of Medicine—six doctors, five surgeons and three apothecaries—surrounded the camp bed on which Louis lay, agitatedly chattering in an almost constant cackle. Courtiers who had *entrée* stood behind them, watching what was going on out of morbid curiosity and self-interest—the death of the King would alter their station in life, and not necessarily for the best. It was a very costly vigil, as fifty of them soon came down with the same disease that Louis suffered from and ten died. One man who merely poked his nose in a door paid horribly for his curiosity.

But the nature of the disease still eluded the Faculty of Medicine. Louis was beside himself. He had long suffered from a pathological fear of death and dying. The King, who as a little boy tortured kittens to death, was absolutely terrified that he might be seriously ill. He ordered each member of the Faculty to take his pulse and look at his tongue, hoping that one would recognize what ailed him and know how to save him.

The fluttering physicians decided a second bleeding was necessary, and if that did not work, a third.

"Then I am gravely ill!" shrieked Louis. "A third bleeding will leave me seriously weak. Can it be avoided?" "No," the Faculty of Medicine assured him.

Louis's children entered the bedroom at their father's request, but after they had stood around him for half an hour without his speaking a word, they left. The only time Louis now responded to the outside world was to shout after any doctor who tried to leave the room, ordering the man back to his bedside "as if he imagined that, surrounded by so many satellites, no harm could happen to his Majesty."

The situation took a turn for the worse when one of the servants brought a lantern near the King and all around him could see red marks on his face. The doctors gave each other astonished looks. The King was supposed to have had smallpox back in 1728, but it certainly looked like the same disease again.

A note was slipped to one of the princesses. "Good Lord," she shouted, "the King has smallpox!"

The male heirs and their wives were ordered from the sickroom, but the Mesdames refused to budge. Although none had had smallpox, they chose to stay at great risk to themselves. All of them eventually came down with the disease but none died.

Everyone was assured that men of Louis's age frequently survived the disease, but not all were convinced. By the next morning the sores were so close to each other that the disease could be identified as confluent smallpox, and Louis's days obviously were numbered.

Those in the room decided not to tell Louis he had smallpox. But the King, with his terror of death, knew almost as much as his doctors and he soon suspected.

"Were it not that I have had the smallpox I should believe that I am about to have it," he said. The Faculty of Medicine averted their eyes.

Meanwhile word of the nature of the illness spread throughout Paris. In 1744, the first time the King was seriously ill, some six thousand Masses were said in Notre Dame. In 1757, when he was sick again, the number fell to six hundred. Now prayers were ordered again for the King's recovery and three people showed up.

Many more, however, crowded the anteroom and the bedchamber. A few, mostly old retainers and his children, honestly cared for Louis; the rest postured and primped and thought about the next Louis while the current one lay terrified in bed.

At night, however, the room was cleared and Du Barry was let in through the private entrance. All night long every night she watched over the King, wiping his brow, chatting with him when he was awake. She too lacked any immunity from smallpox and her courage bespoke honest affection for her dying lover. (She did not catch the disease.)

But the state of the King's soul—and the fate of his mistress—now became the center of a dispute that raged around the royal bed and in the halls. By a peculiar inversion, it was the religious faction among the courtiers that fought against bringing in priests, fearing that Communion would so frighten Louis as to kill him. The nonreligious thought it was time that Louis was prepared to meet his Maker. On another level, of course, the dispute was between the allies and the foes of the royal mistress.

Du Barry's enemies planned a mass invasion of the sickroom to get the King's attention, but they finally agreed to convince the reluctant Archbishop of Paris to make an appearance and demand the sacrament.

That threat sent Du Barry and her allies into a council of war. It was decided to make sure that the Archbishop was never alone with the King and that all their conversations were overheard. The next day, Monday, the Archbishop, a timid man who greatly feared making any enemies at court (what if the King survived?), hesitantly pushed his way through the throngs in the anteroom but was immediately intercepted by one of Louis's ancient cronies, the aging satyr, the Duc de Richelieu.

Richelieu, seventy-eight years old, a white-thatched, red-nosed man who embodied everything immoral and licentious about the court of Louis XV, was quite prepared to defend the King and himself.

He pulled the Archbishop aside and sat with him on a bench in a side room. What was said is not known, but it appears that he convinced the prelate that the dismissal of

Du Barry was too dangerous for the Church. Finally the Archbishop, his resolve gone, entered the bedchamber, where he saw Du Barry perched on the royal bed. At the sight of him the woman fled.

The Archbishop asked about Louis's health and left, his assigned task still not done. Louis, who had expected the Archbishop to bring up the subject of Communion, mistook his speedy departure as a sign that he was not so seriously ill after all. He called Du Barry back to his bedside, "wept with joy" and covered her hands with kisses.

The anti-Du Barry faction tried another prelate, the Cardinal de la Roche-Aymon, but he wiggled out of his predicament by whispering his conversation with the King and no one ever knew what he said and nothing came of it.

The next day there was a great improvement in Louis's condition and the anteroom was crowded anew by courtiers who had assumed earlier that the King was at death's door. That night, however, the disease renewed itself with vigor and Louis's condition deteriorated. He was sometimes delirious.

Worse, the following morning Louis looked at his hands and saw the pustules. He was now sure what it was and cried out in terror, "I have the smallpox."

"At my age," he added when he was calm enough, "one does not recover from this disease. I must put my affairs in order."

So Louis took matters into his own hands. He sent for Du Barry.

"Madame," he told the weeping woman, "I am very ill. I know what I must do. We must part. Go to Rueil, to the Duc d'Aiguillon's château; await my orders there, and be assured that I shall always entertain for you the most tender affection."

Du Barry was crushed, but she knew that Rueil was only

about 12 miles (19 kilometers) away and she could return in a hurry if Louis sent for her. At 4 o'clock the following afternoon the last of the royal mistresses of France left Versailles, never to return.

Still the King hesitated. Later that night he called to his valet and asked him to bring Du Barry.

"Sire, she has gone," the man said.

"Gone, where?" the King asked.

"To Rueil, Sire."

"Ah. Already," the King sighed. "Gone, as we must all go."

The next night the King slipped further. At 5 a.m. he asked that his confessor, the Abbé Maudoux, be brought to him. After a brief meeting with the King, it was announced that the sacrament would be administered at 6 a.m. Louis asked that his grandchildren be awakened so that they could watch.

The Swiss Guards and the Royal Bodyguard lined up in the vast glittering hallways. The royal and holy procession, carrying torches and wearing formal robes, followed by almost the entire court, marched slowly to the King's chambers.

Louis was now in the last stages of the disease. His face was black and swollen. His sores gave off such a stench that not even open windows, incense and medicine could overcome it. He was literally rotting to death. Weakly, he tried to rise.

"If God pays the honor of a visit to such a wretch as myself, the least I can do is to receive Him with respect," Louis said.

One of his daughters, who had been so shocked by her father's behavior that she had become a nun, had lovingly sent Louis a crucifix. He clutched it now as the Latin prayers were recited. The Cardinal de la Roche-Aymon, in his purple robes, bent over the dying man and whispered the words of the Holy Communion service. Just as he turned away the Abbé clutched his sleeve and said something into the Cardinal's ear. The Cardinal nodded and then walked to the door and repeated the conditions under which the sacrament would be communicated.

"Messieurs," the Cardinal said, "the King charges me to inform you that he asks pardon of God for having offended Him and for the scandal he has given his people, that if God restores him to health, he will occupy himself with the maintenance of religion and the welfare of his people."

Only two people spoke afterward. Richelieu directed an obscenity at the Cardinal. The King murmured, "I should have wished for sufficient strength to say it myself."

That ended the ceremony, and the anteroom emptied suddenly as the courtiers and the family fled from the pestilence and the incredible stench of a smallpox death.

Louis's condition continued to sink. The next day he received Extreme Unction and the death agonies began.

At 3:15 the following afternoon the Grand Chamberlain appeared at the door of the main courtyard of Versailles and proclaimed: *"Messieurs, le Roi est mort! Vive le Roi!"*

As tradition required, the First Gentleman of the Bedchamber told the First Surgeon to open and embalm the body. To do so would have been a death sentence to the surgeon.

"I am ready," was the doctor's wily reply, "but while I operate you must hold the head as your duties command."

So no autopsy or embalming was performed. The great-grandson of the Sun King was dumped unceremoniously into a lead-lined casket. Alcohol was poured into the coffin and the remains soaked with quicklime. On the night of May 12, 1774, Louis XV was taken to Saint-Denis cemetery, followed only by one friend and a hundred guards.

V

La Noche Triste

Shortly after Hernando Cortes had marched for the second time into the silvery island city of Tenochtitlán in June, 1520, he had put Montezuma, the Aztec emperor, under house arrest. To the weak and despondent Indian, who had reigned over a vast and obedient empire, this was the final insult. He sank into a sullen depression.

Once, before yielding to Cortes' threats, he had summoned all his lords together and tearfully urged them to accept the Spaniard's domination and to produce the gold that the Spaniards said they needed. His lords came up with a large treasure which was divided among the emperor, Cortes and Cortes' few hundred soldiers. But while the captain-general had been away a hideous massacre had been ordered by his second-in-command, Pedro de Alvarado. The massacre, made by ambush, shocked the Aztecs, who never knew warfare could be fought in such a cowardly fashion. Cortes had to race back across the drawbridges that spanned the lake to relieve his surrounded garrison in the Aztec capital.

Now, two weeks later, with food and water running out and Aztec assaults beginning to breach his last defenses, Cortes had to try to get his army off the island in the lake. The Aztecs were now under the leadership of Montezuma's brother, Cuitlahuac, far stronger and abler than the doomed emperor, and a week of violent clashes had placed the Spaniards on the edge of defeat. Only the terror of what would befall them if they were captured by the Aztecs maintained them. They had seen the bloody sacrifices on the stone altars and knew that the Indians were quite capable of organizing cardiectomies by the hundreds or thousands.

Cortes, who had started the conquest of Mexico months earlier with 550 soldiers, sixteen horses, ten brass cannon, four falconets and sixteen arquebuses, was about to begin his *Noche Triste,* the Night of Sorrows.

After the Indians cut off the regular spans to the mainland with their canoes, he had constructed a portable bridge. His men loaded what treasure they could on their horses. Those soldiers who were newcomers from the forces in Cuba tried to carry their shares themselves, but Cortes' veterans knew better. The Spaniards waited until the moon went behind a cloud, and then, as a terrible thunderstorm broke over Tenochtitlán, they quietly left the fortress.

The strong refused to support the weak. The wounded and the sick, including one man with smallpox, fended for themselves.

Cortes' army reached the first causeway before the alarm went up. The priests of Huitzilopochtli began beating the drums from the top of the pyramid. Aztecs swarmed along the causeways and bridges.

Burdened by their treasures, many of the Spaniards and

their Indian Tlaxcalteca allies could not return the fire and began falling. In a rush the Spaniards had jammed the portable bridge and now were trapped in the open.

Canoes converged on the causeway, arrows and spears arched through the dark, turbulent skies. The Spaniards fired back, but a hundred Indians appeared where one fell.

Cortes' allies suffered the most and began pushing from the rear. The front line was trampled into the twelve-foot-deep canal. Horses and baggage fell on top of them, followed by still more bodies. The Spaniards found that they were able to escape on the bridge made by the dead. They repeated the grisly crossing again and again.

The treasure soon disappeared into the fetid water of the lake. So did the artillery and most of the horses. Few of the sick or wounded survived.

A priest, Francisco de Aguilar, writing years later, remembered:

> As we were now fleeing it was heartbreaking to see our companions dying, and to see how the Indians carried them off to tear them to pieces. The number of Indians pursuing us could have been five or six thousand, because the rest of the horde of warriors were occupied in looting the baggage that had sunk in the canals. They were even cutting [one] another's hands off to get a larger amount of the plunder.

But with Cortes routed, Cuitlahuac believed that he had plenty of time. He ordered those soldiers still obeying orders to get as many of the Tlaxcaltecas as they could for sacrifice to satisfy the gods. He felt that now the Spaniards could be defeated at any time.

But Cortes had left a time bomb behind him: the man with smallpox.

Aztecs looted all the dead they could find, probably turning over the smallpox victim—Spaniard, Tlaxcalteca or Cuban, no one knows for sure—and inhaling the viruses that were streaming from the victim.

Within two weeks smallpox was ravaging Tenochtitlán and the area around the lake, wiping out perhaps one quarter of the population. The Indians had no natural defenses against the virus, because the disease had not existed in the Americas until the Spaniards brought it on their small, crude ships. It rampaged through the virgin population. By the time it had finally run its course, the Aztec army had been greatly reduced and Cuitlahuac himself, along with many other leaders, was dead of pox. The disease had taken the one man who might successfully have led the assault against the weakened Spaniards. When Cortes returned to the attack on Tenochtitlán a few months later, it was the Aztecs who were the most weakened and he defeated them easily. After the battle he found so many dead lying about the once-great fortress city that he claimed that he could not walk through the streets without stepping on a body. The Aztec empire was no more.

Variola was now loose in the land. In two years between three and four million Indians would be dead of smallpox, and many other empires would have been led into bondage.

The first infections in the New World probably came from slave ships. There is no indication of a disease similar to smallpox in any lore of the New World before 1492.

By 1519 smallpox had already spread across the Caribbean, jumping from Santo Domingo to Puerto Rico.

Slave ships had been making the run from Africa regularly since 1510, and their cargoes almost certainly included bearers of smallpox. The disease was endemic to Africa by then.

Slave ships were ideal incubators of smallpox and other diseases. The vessels spent months off the African coast loading because it was uneconomical for them to cross the ocean with anything less than a full load. Even after they had crammed in all the slaves they could hold, they had to store supplies for the long voyage. Consequently it was entirely possible for a captured African to spend a hundred days chained in his six-foot-by-one-foot floating cesspool.

A slave incubating any disease would quickly pass it on to his shipmates. If the crew discovered anyone with smallpox they routinely killed him. The goal was to try to land with a cargo that appeared completely healthy; however, some ships arrived in the New World with their cargo infested with smallpox.

Many historians claim that the actual death toll from smallpox on those ships was not very great. But the effect on the Indians and colonists was catastrophic.

After the ship landed, the slaves and their white masters freely mingled with the natives. It soon became possible to trace epidemics of smallpox from an individual slave ship in port.

The Indians, in the words of one missionary, "die so easily that the bare look and smell of a Spaniard causes them to give up the ghost." Millions gave up the ghost.

Their social structure heightened their natural susceptibility. The Indians were naturally gregarious within their tribes, living in close quarters with one another. The healthy in one hut could not avoid the sick in the hut a few feet away. And when a tribe was nomadic, it could carry the disease to all neighboring tribes as it followed its ageless trek across the plains and mountains of America.

This natural susceptibility persisted. As late as the nineteenth century Indians were suffering 55–90 percent mortality directly or indirectly in smallpox epidemics, compared to 10 percent or less of the white population.

The Aztecs were only the first to succumb to the disease. Next to suffer were the bellicose tribes of the Central American highlands in what is now Guatemala and Honduras.

Three tribes lived in that area: the Quiche in the west, the Cakchiquel in the east, and the Zutugil between them. Rivalry among the three tribes was intense and warfare was frequent.

In what was to be the final round of this bloody rivalry, the Cakchiquel finally subdued the Zutugil in 1501 and made an alliance with the as yet unconquered Aztecs. Feeling their new power, predictably they turned on the Quiche, but nature, this time in the form of a plague of locusts, intervened and the Cakchiquel suffered terribly.

Around 1519 the war was suspended because of recurring rumors of tall white men with beards and great wooden ships. The tall white men were the Spaniards, led by Alvarado, who were moving south from Mexico. Ahead of him was a wave of smallpox.

Tribe after tribe was stricken as he moved south. As a result, Alvarado and his men conquered them easily. The disease finally spread from the Nahua to the Quiche and Cakchiquel, destroying almost half their populations. But animosity dies hard. What was left of the Quiche empire believed that it was still stronger than the Cakchiquel remnant, and they attacked just one more time. The Cakchiquel ran to Alvarado for assistance. Always the scoundrel, Alvarado captured both tribes and two empires fell to variola and the Spaniards.

The disease kept spreading southward, down the isthmus of Panama through Colombia to Peru. The next empire to fall was the mighty Inca.

The Incan civilization had reached its zenith and internal decay had already set in. It controlled an area about the size of the U.S. eastern seaboard and was ruled from Cuzco by the last of the mighty emperors, Huayna Capac.

To ease administration, Huayna Capac decided to divide his empire into two provinces, with one son reigning in

Quito and another in Cuzco. His heir as Great Inca was his son Niña Cuyoche.

To help secure his domain Huayna moved his professional army to the northernmost section of the empire sometime around 1525. It was during this campaign that he first heard of the arrival of the Spaniards.

At that moment the Incas were hit with a disastrous smallpox epidemic. The disease apparently had crossed the mountains on foot rather than in the ships of Francisco Pizarro. Pizarro was at sea for too long and there is no record in his logs of smallpox. However when it came, between 1525 and 1527, the power structure of the Incas was almost destroyed. Among the dead were Huayna Capac, his brother, sister, uncle and Niña Cuyoche. The forces that held the empire had been sliced down.

Civil war broke out between the surviving sons, Atahualpa and Huáscar. It was exactly at that time that Pizarro arrived.

Pizarro was raised as a swineherd and behaved like one for all his adult life. He cheated, stole, tortured, lied, killed and swindled. Along with him came such luminaries as Hernando de Soto, Nuñez de Balboa, and Pizarro's brother Hernando, none of whom was very much better.

With a contingent of sixty-two horsemen, a hundred and six foot soldiers and a few cannon, the Spaniards destroyed four centuries of Incan civilization in a few years. Smallpox had done their work for them.

Characteristically the decisive battle was begun with an ambush in which the Spaniards trapped and slaughtered several thousand defenseless Incans carrying Atahualpa, who had earlier defeated Huáscar, on a golden throne. Atahualpa was captured. He was executed finally by Pizarro, forever ending the Inca dynasty. (Just before Atahualpa was to die by fire he was advised that things would go better if he converted on the spot to Catholicism. He agreed, converted and was promptly garroted.) By 1535 the empire was gone. Pizarro was murdered in 1541.

Legend lauds the conquistadores, but actually most of them were uncivilized brutes whose barbarities were sanctified by fanatic priests. Their conquest of the Americas was less a tribute to their courage than to the virulence of variola. Smallpox conquered the Aztec and the Incans, the Quiche and the Cakchiquel. The Spaniards were merely the carriers of the virus, who cleaned up when the virus had done its work.

The effects of the disease on the Indian population cannot be exaggerated. Today, in the two provinces of the Incas, there are about four thousand people, about one tenth the size of the small army that accompanied Atahualpa into his ambush.

In forty years the town of Chincha Alta, south of Lima, lost four fifths of its population. Ten years later only a few Indians were left. Spaniards reported that in the valleys of Lunahuana and Huarcu the population had fallen from thirty thousand to two thousand by 1600. Another wrote:

> We cannot conceal the great paradox that a barbarous Huayna Capac kept such excellent order that the entire country was calm and all were nourished, whereas today we see only infinite deserted villages on all the roads of the kingdom.

A number of the honorable Spaniards in the New World—and there were many—urged the king to restore some semblance of order to the lands of the Incans before all the Indians were killed off. One implored:

> I must advise Your Catholic Majesty that the wretched Indians are being consumed and are dying

out. Half have disappeared, and all will come to an end within eight years unless the situation is remedied.

There was very little the King could do. It is estimated that the Spaniards and smallpox killed enough Indians to decrease the native population of Peru to one quarter of its pre-Pizarro size.

Smallpox, of course, was not the only disease that ravaged the Indians, nor were the Spaniards the only carriers. Sir Francis Drake and his English fleet brought typhus in 1585. Influenza, then epidemic in Europe, also made the ocean voyage. Apparently the Indians were hit with successive waves of disease, one deadly type after another, and none provided immunity against the ones that followed.

Smallpox was still the most terrible. Between 1585 and 1591, just after the population of Peru had begun to stabilize, thirty thousand died in Quito alone. One quarter of the population of Lima also fell despite the rare humanity of Viceroy Don Ferdinando de Torres y Portugal, who set up quarantine measures with the help of Spanish physicians.

By 1560 the disease had spread to Chile.

What the viruses did not kill succumbed to the appetites of the conquerors. Terror reigned in Spanish America. "Unless orders are given to reduce the confusion in the goverment of this land," one Spaniard wrote to the king, "its natives will come to an end; and once they are finished Your Majesty's rule over it will cease."

The Spaniards, incidentally, could not blame the Indians for the filth that helped spread diseases. A number of Spaniards in the early years marveled about the cleanliness of the Indian villages, especially compared to the towns of Spain. When the villages became filthy it appears that the Spaniards were responsible.

The Spaniards were not the only disease carriers. The French discovered the magnificent port of what is now Rio de Janeiro in 1555 and named it France Antarctique. Within ten years they were driven off by the Portuguese. The main contribution the French made to the area was a smallpox epidemic that killed off half the natives of what is now the state of Bahia. With the work of the slavers and the demoralization of the conquered natives, soon there were no more Indians on the southern coast of Brazil.

By the early seventeenth century the English were importing slaves into their North American colonies and with them, apparently, came smallpox. The first recorded epidemic to strike the Indians was in 1633, when the natives surrounding Boston were virtually wiped out. John Winthrop wrote that in one day the Indians buried thirty of their tribesmen. By the first month of 1634, seven hundred Narragansetts and nine hundred to a thousand Connecticuts had died of smallpox. Winthrop noted that "many did rott above ground for want of burial." The epidemic was so severe that the Dutch, who had been trying to trade with the Indians of New England, beat a hasty retreat back to New Amsterdam to avoid the disease.

From 1633 to 1641 smallpox ravaged the Great Lakes region, extending into Canada. In 1639 the first Indian hospital, l'Hôtel-Dieu, was established in Quebec by French nuns. No sooner did it open than a smallpox epidemic broke out. More than four hundred children alone died from the disease, and the mother superior wrote:

> I have seen many whose bodies are entirely covered with smallpox and in a burning fever. . . . The remedies that we brought from Europe are very good for the savages, who have no difficulty in taking our medicine, nor in having themselves bled.

During an epidemic among the Hurons in 1640 the Indians noticed that Christian Indians seemed to suffer more than the nonconverted. They quickly, and properly, deduced that the missionaries, who sometimes volunteered to treat the sick as part of their Christian calling, were spreading the disease as they went from tribe to tribe. They shut off access to many of the missionaries as a result. The Indians were equally responsible for spreading smallpox; the Hurons, for example, infected the Algonquians.

In 1650 another epidemic hit Quebec and l'Hôtel-Dieu was again filled. The nuns built a fence to extend the wards beyond the original buildings in an effort to hold all the victims.

Also in about 1640 smallpox struck the Iroquois nation. The Mohawk tribe of the Iroquois was afflicted in 1669. The Mohawks were then in the middle of a great war with the New England tribes. The conflict ended in a great battle in which several of the combatants were suffering from smallpox. The ensuing epidemic left no victors.

That was not the only Indian war terminated by smallpox. In 1690 a union of English, Mohegans and Iroquois attacked the French at Quebec, and the Count de Frontenac wrote that "the affair would have been embarrassing had not God interposed." The English and the Mohegans were attacked by smallpox just before the three allied armies were to rendezvous. The Iroquois saw the pox marks on the faces of their allies and ran. It was, unfortunately, too late. Three to four hundred Indians in the two tribes died. That ended the alliance. When the English again asked for help from the Iroquois, they were refused.

Meanwhile, the disease had spread into the southern colonies. In 1667 sailors landed in Virginia and passed smallpox on to the Indians. One correspondent wrote: "Practically every tribe fell into the hands of the grim reaper and disappeared, the only exception being the Gingaskins." The tribes also spread the disease even farther south into Georgia and the lower Mississippi.

In 1696 the Quapaws were smitten and only a few survived. In one village every child died.

In order to keep the Indians from passing the disease back to the colonists, a number of towns barred Indians that might be infected. On March 2, 1662, the town council of East Hampton, Long Island, ruled:

> It is ordered that no Indian shall come to towne into the street after sufficient notice upon penalty of 5 s, or to be whipped, until they be free of the small poxe; but that they may come where they have corner on the back side and call; and if any English or Indian servant shall go to their wigwam they shall suffer the same punishment.

Indians were allowed in the backdoor only.

In 1708 l'Hôtel-Dieu was again jammed with smallpox victims. A nun wrote:

> Each day the bodies of the dead are conveyed into the church in the lower town, or into the cathedral without any ceremony, and in the evening they are buried together, sometimes as many as 15, 16 or 18—and this lasted for many months.

The dead at l'Hôtel-Dieu were buried in a cemetery called *Cime tieres des Picotées.* In 1854 the cemetery was dug up for some water and sewer pipes, and an outbreak of smallpox, centering around the grounds, struck Quebec. The coincidence was duly noted by newspapers.

By 1709 the disease reached the Indians of California and many of the children in the Spanish missions died. It was endemic in California by 1728.

In 1717 the Iroquois were hit again. The Governor of New York explained the disease to them:

> . . . We Christians look upon that disease and others of that kind as a punishment for our misdeeds and sin, such as breaking of covenants and promises, murders and robbery and the like—for in Pennsylvania . . . there is not one Christian family that has escaped the disease, and at present it rages and has for 12 months past in the Jerseys. . . . And we firmly believe yet so long as we continue the practice of these sins, our plague will also continue.

In one way the Indians made their situation worse than it needed to be. Their treatment was the exact opposite of the European pre-Sydenham method but just as dangerous. First they put the sick person in a sweathouse and then forced him to jump into ice-cold water. It was the Indian version of the Scandinavian sauna, but when applied to someone with smallpox it was frequently fatal.

The Indians thought the cure made as much sense as the white man's explanation of retributive disease. In fact, there was some relation. One Indian explained to a trapper:

> Our debauches and fatigues create depraved humors, which nature would throw out of the body if she had but strength enough to open the gates, viz. the pores of the skin. 'Tis true she expels as much as she can, by urine and stool, by the mouth, nose and insensible transpiration. But sometimes the quantity of the serosities [*sic*] is so overbearing that they overflow all parts of the body between the skin and the flesh; and in that case 'tis our business to procure their egress the speediest and shortest way, for fear their longer stay should give rise to this gout, rheumatism, dropsy, palsy and all the other distempers that sink a healthy state. Now, to compass this end, we must unlock the pores by the means of sweating; and withal take care to shut

> them soon after lest the nutritive juice should glide out by the same passage; which can be no otherwise prevented than by throwing ourselves into cold water, as we usually do.

Indian theology and etiology were not much different from those of the colonists. In 1739 a slave ship landed at the port of Charleston and contaminated slaves on board soon spread the disease to the Cherokees. The Cherokees tried treating the disease with the cold-water treatment, and when that failed to halt the epidemic, the medicine men told them that they were being punished for their sins, particularly adultery among the young married couples. Sinners who caught the disease were forced to lie outside day and night and to pour cold water on their bodies to reduce their fever. It may have worked but it was very unpleasant, and despite the considerable amounts of chanting, singing and praying accompanying the treatment, a number of Cherokees committed suicide rather than continue with a cure.

The first known attempt at quarantine in North America was around 1730, when a trader convinced the Muskogees to stay away from other tribes which had the disease.

Mexico continued to suffer mightily. It was hit again in 1747, 1763 and 1779. In the last of those outbreaks a journalist reported that in the fifty-seven days between October 22 and December 18, there were 44,286 cases with 8,821 of them fatal, a mortality rate of 20 percent.

The largest epidemic in North America began in 1752 and spread through all the tribes. There were some instances in which the whites helped the disease along to clear the Indians off land they wanted. One letter written by a M. de Longueuil says, "'Twere desirable that it [smallpox] should break out and spread generally throughout the localities inhabited by our rebels [Indians]. It would be fully as good as an Army."

Warfare continued to be the prime method of spreading epidemics. During the French and Indian Wars, Montcalm besieged the English garrison at Fort William Henry on Lake George. Disease forced the English to surrender to the French and their Indian allies. The Indians rushed in and butchered the sick and dug up the corpses in the cemetery to take scalps.

Many of the sick and the dead had been smallpox victims, and the Indians came down with the disease. The French were furious at the barbarism and the French government was furious with Montcalm for letting it occur. One letter from Versailles asked, "How could it happen that the smallpox among the Indians cost the king a million francs? What does this expense mean?"

Occasionally smallpox was used as a weapon in biological warfare. In 1763, during an Indian uprising near Fort Pitt, a friend of the British commander, Sir Jeffrey Amherst, decided to upset the normal military equation. He wrote to Amherst: "I will try to inoculate the [Indians] with some blankets that may fall into their hands, and take care not to get the disease myself."

Amherst thought the idea was fine. "You will do well to try to inoculate the Indians by means of blankets as well as to try every other method that can serve to extirpate this execrable race."

Shortly thereafter an officer with the British wrote in his diary: "Out of our regard for them [two Indian chiefs] we gave them two blankets and a handkerchief out of the smallpox hospital. I hope it will have the desired effect."

It did. Within a few months hundreds of Mingoes, Delawares and Shawnees were dead.

On the other hand, several planned massacres of whites

were prevented by smallpox. In 1778 the Indians in the Hudson Bay area of Canada went on the warpath after a trader murdered one of them. But before they could get a real massacre going, smallpox struck and two thirds of the Indians died. A trader in the vicinity wrote that corpses were lying everywhere, unburied, eaten by their own dogs and by wolves. Other massacres in the northwest were also averted when the disease disabled warring Indian tribes. The Cree and Chipewyan suffered the most.

Sometimes the disease seemed to have a justice all its own. In 1781 an army of Kenistenos, Assiniboin and Ojibways attacked a defenseless Gros Ventres village. The Gros Ventres were suffering from a terrible outbreak of smallpox and the attackers found resistance weak and a village already stacked with the dead. The stench was so terrible that they could stay only to take a few scalps. On the return trip to their villages the attackers began dropping one by one. Only four Indians eventually survived. Those four communicated the disease to the rest of their tribes, who, in turn, fled and spread the disease even further.

Between fifteen hundred and two thousand Ojibways died in this contagion.

The disease ranged over the entire continent. While the Ojibways were dying, smallpox hit the southwestern pueblos, killing five thousand in New Mexico. In Alaska three epidemics struck, incluuing a major outbreak in 1787.

Back in Mexico the Spaniards were still trying to prevent epidemics. The viceroy issued strict quarantine and health measures in an effort to limit an epidemic in 1797. Victims were ordered to segregate, whole towns were quarantined and mail was fumigated with sulfur. But by this time enough adults either had been killed off or had acquired immunity for the disease to begin settling down to a childhood affliction. Most of the victims in the 1797 outbreak were under the age of twenty. The viceroy's measures did not work, because the Indians, ever distrustful of Spanish authority (and who had better reason?) refused to obey and hid victims where the viceroy's agents couldn't find them.

Few attempts were made to inoculate the Indians in the Americas, and when it was tried, variolation* frequently killed the subjects and touched off independent epidemics. The 1797 Mexican-Guatemalan epidemic was the first in which any real attempt was made to variolate the natives. In the city of Valladolid, sixty-eight hundred Indians were inoculated and a hundred and seventy of them (2.5 percent) died. The clergy seemed to be the main proponents of inoculation, and in parts of southern Mexico particularly around Oaxaca and Tehuantepec, they did well. Some sixty to seventy thousand people were variolated. Often results were not as striking. One priest lamented in 1796:

> These people are the most stubborn in the world. No reasons are sufficient to convince them of the benefits of inoculation. Some say that God sent the disease to the town but they will not permit the Spaniards to give it to any more of their children.

The price the Indians paid for the white man's colonization was bitter and ghastly. Not the least of the burdens was smallpox.

---

* Variolation means actually infecting a patient with smallpox virus in the hope he or she will get a mild case and therefore acquire life-long immunity.

VI

Jenner

In 1757 a smallpox epidemic broke out in Gloucestershire, England.

Church attendance dropped to nearly nothing. Schools closed, and the children who boarded in them were sent home. Village social life ended. The sickness raged from house to house, up and down the gentle green hills. As usual, it attacked mostly the very young.

In order to protect the children who so far had escaped the disease, a number of parents decided on a dangerous, but frequently effective course. They would "buy the pox."

It had been known for centuries that once someone contracted smallpox and survived somehow he was protected from the disease for the rest of his life. The phenomenon was, of course, unexplained. What was known was that—with very rare exceptions—no one ever got smallpox twice. Survivors could nurse the sick who might be in the worst stages of confluent smallpox—hold them, soothe them, wash them—yet whatever it was that caused the disease could do no harm to the nurse.

There were a number of explanatory theories. One, stated by a Belgian physician early in the seventeenth century, suggested that:

> . . . after the organs which secrete this poison have felt its tyranny, so great an aversion to the horror is conceived, that great precautions against its reproduction are taken; lest from carelessness they should fall into the same evil.

It was inevitable that somebody should decide that by deliberately incurring a mild case of smallpox—one that was survivable and not greatly disfiguring—protection against a more serious case could be acquired. That was called "buying the pox."

Some children of Gloucestershire who so far had escaped the disease were taken to a stable in Wooton-under-Edge owned by a Mr. Holbrow, an apothecary. They had been prepared for this visit for six weeks. They were bled repeatedly to make sure their blood was "fine" and were administered violent purges to empty their stomachs. Sometimes they were given a sickeningly sweet herb tea, but they were otherwise starved, almost to the point of serious harm. When they finally arrived at Mr. Holbrow's stable they were emaciated, almost unable to walk by themselves and terribly ill, hardly prepared for the ordeal that would follow.

One of those children was named Edward Jenner.

Like the other children, Jenner, aged eight, was put up on a table where Holbrow scratched his arm several times with the tip of a knife, placed the dried scab of a smallpox victim over the fresh cuts in young Jenner's arm and bandaged the arm.

Jenner was forced to remain in the stable for several weeks, surrounded by others who were in various stages of the illness. After about a week he too came down with a mild case of smallpox. He broke out into a fever and the typical bright red rash. Three days later the symptoms cleared up, but it took almost a month for the boy to recover fully from the combination of smallpox and the torture that had preceded the inoculation.

Jenner was now immune, but the horror of what he went through that summer in 1757 stayed with him throughout his life. Perhaps that horror became part of his later stubbornness.

The practice of variolation had begun in the Orient at an unknown time. The Chinese made their children inhale the dried scabs of smallpox sores from straws made of ivory. In Africa and the Near East scabrous matter was inserted into cuts.

Teams of traveling Brahmans variolated in parts of India, and even in the remoter parts of Africa variolation was common.

The first notice in the West was in Denmark, where the physician Thomas Bartholin wrote in 1675 that peasants in his country were practicing variolation. Soon physicians in England were reporting that it had been going on there for centuries.

It was an ancient custom, according to Dr. Perrot Williams in the Royal Society's *Philosophical Transactions* in 1722. In Pembrokeshire he found:

> In order to procure the distemper to themselves, they rub the matter taken from the pustules, when ripe, on several parts of the skin of the arms, etc., or prick those parts with pins or the like, first infected with the same matter. And notwithstanding they omit the necessary evacuations, such as purging, etc., yet, enough; and

> what's remarkable, I cannot hear of one instance of their having the Small Pox a second time.
>
> There are now living in this Town [Haverford West] and neighbourhood five or six persons, who undoubtedly had that distemper after taking the foresaid method to infect themselves: one of whom, a young woman aged twenty-three, told me . . . that, about eight or nine years ago, in order to infect herself, she had twenty pocky scabs (taken from one toward the latter end of the distemper) in the hollow of her hand a considerable time; that about ten or twelve days afterwards she sicken'd, and had upwards of thirty large pustules in her Face and other parts; and that she has since freely conversed with such as have had the Small Pox on them.

Nevertheless, the practice was still not widely recognized by physicians in the West. The person who brought it to respectability in England and Western Europe was Mary Pierrepont, a beautiful young woman, full of high spirits and of irrepressible intelligence. She was one of the typical products of the more enlightened views that had been sweeping Europe since the last century—willing to accept adventure, go to any lengths to satisfy her curiosity, ignore superstition. "I came very young into the hurry of the world," she wrote in one of her famed letters.

In 1713 her brother died of smallpox and two years later, at the age of twenty-six, she too caught the disease. It left her complexion badly scarred and destroyed one of her eyebrows.

When her father would not agree to let her marry the man she loved, Edward Montagu, she carried on a two-year romance by letter and finally eloped with him in 1714 into what, sadly, became an unhappy marriage.

In 1716 Montagu was made ambassador to Turkey and the new Lady Mary Wortley Montagu accompanied him to Constantinople. She wrote a torrent of letters describing her trip across Europe and life in the city where European civilization confronted the Orient. Because she was the wife of an ambassador few doors were closed to her; because of her great intelligence few things escaped her. She breached even the security of the sultan's women's quarters to see what was going on behind the veils. Her letters became renowned in eighteenth-century England, making her a woman of considerable power and influence in literary circles.

It was in Constantinople that she first came upon the practice of variolation. On April 1, 1717, she wrote a friend, Sarah Chiswell from Adrianople, first describing the outbreak of the bubonic plague in the area and assuring her friend that "those dreadful stories you have heard of the plague have very little foundation in truth." Dropping that subject, she added:

> Apropos of distempers, I am going to tell you a thing that I am sure will make you wish yourself here. The smallpox, so fatal and so general amongst us, is here entirely harmless by the invention of engrafting (which is the term they give it). There is a set of old women who make it their business to perform the operation. Every autumn, in the month of September, when the great heat is abated, people send to one another to know if any of their family has a mind to have smallpox. They make parties for this purpose, and when they are met (commonly fifteen or sixteen together) the old woman comes with a nutshell full of the matter of the best sort of smallpox and asks what veins you please to have opened. She immediately rips open that you offer to her with a large needle (which gives you no more

pain than a common scratch) and puts into the vein as much venom as can lie upon the head of her needle, and after binds up the little wound with a hollow bit of shell, and in this manner opens four or five veins.

The Grecians have commonly the superstition of opening one in the middle of the forehead, in each arm, and on the breast to mark the sign of the cross, but this has a very ill effect, all these wounds leaving little scars, and is not done by those that are not superstitious, who choose to have them in the legs or that part of the arm that is concealed. The children or young patients play together all the rest of the day and are in perfect health till the eighth. Then the fever begins to seize 'em and they keep to their beds two days, very seldom three. They have very rarely above twenty or thirty in their faces, which never mark, and in about eight days' time they are as well as before their illness. Where they are wounded there remain running sores during the distemper, which I don't doubt is a great relief to it.

Every year thousands undergo this operation and the French ambassador says pleasantly that they take the smallpox here by way of diversion as they take the waters of other countries. There is no example of anyone that has died in it, and you may believe I am very well satisfied of the safety of the experiment since I intend to try it on my dear little son. I am patriot enough to take pains to bring this useful invention into fashion in England, and I should not fail to write to some of our doctors very particularly about it if I knew any one of 'em that I thought had virtue enough to destroy such a considerable branch of their revenue for the good of mankind, but that distemper is too beneficial to them not to expose to all their resentment the hardy wight that should undertake to put an end to it.

Perhaps if I live to return I may, however, have the courage to war with 'em. Upon this occasion, admire the heroism of your friend, etc.

Lady Mary was as good as her word. She consulted with her family physician, Dr. Maitland, who was in Constantinople at the time, and with Dr. Emanuel Timonius, who had written a generally ignored article on variolation for the Royal Society's *Philosophical Transactions* three years earlier.

Without her husband's knowledge, she ordered that her six-year-old son be inoculated. Timonius and Maitland, who had chosen the donor, watched as an old Greek woman prepared for the procedure. She took out a blunt, rusty needle and began to scratch the boy's arm. Her hands shook so much that she put the boy through agony and he began to shriek. Maitland finally took his own instrument and inoculated the child's other arm himself.

A week later Lady Mary informed her husband, assuring him that the boy was "at this time singing and playing and very impatient for his supper. I cannot engraft the girl; her nurse has not had the Small Pox."

The boy survived with no ill effects.

Lady Mary returned to London a few years later. In 1721 an epidemic of smallpox erupted in London and she wanted to protect her daughter. Maitland performed the procedure, which was witnessed by three Royal Society physicians. The society witnesses also visited the girl throughout her convalescence. Newspapers reported the story, and the operation became the topic of considerable discussion in the salons of London.

One of those attracted to the practice was the Princess Caroline. She wanted her children inoculated, but only after a little safety test. Six prisoners who had been condemned to the gallows were chosen and asked to volunteer their services. They were promised freedom if they survived. Five of them came down with slight cases of smallpox after variolation and were set free. The inoculation of the sixth did not take, probably because he had had smallpox as a child but hadn't remembered it. He also was released.

Caroline was still not entirely convinced, so, using her princessly prerogative, she ordered the inoculation of all the orphans in St. James's Parish. When they all survived, Caroline had her two daughters inoculated by Maitland.

Their recovery put the royal seal of approval on variolation and the practice became widespread. But problems soon arose.

Maitland was asked to inoculate the daughter of a Quaker family in Hertford. The disease he induced spread to six of the family's servants, which in turn touched off a local epidemic.

There were two deaths after other inoculations, one a servant in the household of Lord Bathhurst and the other the young son of the Earl of Sunderland. Public reaction was severe.

Enter the clergy. On July 2, 1722, Edmund Massey, the minister at St. Andrews, Holborn, sermonized "Against the Dangerous and Sinful Practice of Inoculation." He cited Job who suffered mightily and nobly, without trying to interfere with God's test of him. Job was afflicted with a loathsome disease "which might be what is now conveyed to men by some such way as that of inoculation which is derived from the same part of the World as was Job's scene of action."

Then, expressing a philosophy that would be improved upon a century later by such eminent social critics as Thomas Malthus, he thundered: "The fear of disease is a happy restraint to men. If men were more healthy, 'tis a great chance they would be less righteous. Let the Atheist and the Scoffer inoculate. Their hope is in and for only this life. Let us bless God for the Afflictions He sends upon us, and grant us patience under them."

There were defenders. One physician, James Jurin, used statistics to support the practice. He reported that of all children born, one in fourteen died of smallpox, while those variolated died at the rate of one in ninety-one. Where he got his figures is not known.

The debate was on. Most of the medical profession sided with the clergy, perhaps because of bad experiences with their patients, perhaps out of snobbery or greed.

Lady Mary was not ready to give up. She worked fervently to get acceptance of variolation, including visiting patients and the physicians who were still practicing it. Statistics bore out her contention that it could be performed safely, but the arguments against variolation, for the time being, prevailed. Lady Mary finally fled the city in order to get some rest from the dispute and to continue with her private life. She had decided to let the scientific community resolve the issue.

The practice of variolation in England waned until 1743, when an American physician, John Kirkpatrick, visited London and told how it had stopped a virulent epidemic in Charleston, South Carolina, five years before. Kirkpatrick had used an improved form of the procedure which had lowered the mortality rate to one in a hundred.

Kirkpatrick revived interest in variolation and the Smallpox and Inoculation Hospital, expressly designed for variolation, was established in 1746. The art was further refined by a father-and-son team, Robert and Daniel Sutton.

In eleven years the Suttons claimed they inoculated 2,514 and made a fortune. They were good enough to have a perfect safety record (or so they claimed) and franchised their secrets for between fifty and a hundred pounds to physicians who were far enough away to pose no competitive threat. Sometimes they would contract for a 50 percent share of the profits.

Their secret was in strict selection of their patients—only the healthy received variolation from them—and unusually attentive and hygienic care in their own hospital.

Some young English physicians were selling themselves as professional variolators, and royal families throughout Europe prided themselves on being inoculated by an English specialist. The family of Louis XV of France quickly had itself inoculated after witnessing his ghastly death.

One variolator, Baron Dimsdale, was called to Russia by Catherine the Great, whose husband had been horribly disfigured by the disease. He inoculated the entire family and received one of the highest fees in history: "a title of nobility, and in money and presents acquired, about 20,000 pounds."

Voltaire pushed the practice in France, and it was sanctioned by the Parlement until 1763, when an epidemic in Paris was traced to a variolation. The Parlement banned variolation within the city limits, but the physicians simply moved to the suburbs and continued.

It was in this context that the young Jenner underwent his ordeal in the inoculation barn in Wooton.

It is fair to say that most of the wealthy and the aristocratic had been protected against the disease by variolation by the time Jenner reached manhood. Except in the areas where variolation was a folk practice, it was the poor and rural who were still frequently unprotected. As long as this continued there was no hope of stopping the disease. Moreover, variolation was still a dangerous procedure, even when practiced by the Suttons or Dimsdale. The patients did in fact suffer from smallpox, and it was not always controllable.

Some efforts had been made to help the poor. Daniel

Sutton agreed to variolate four hundred poor children and care for them for one full month if patrons subscribed a guinea each. There were few takers and Sutton abandoned his plan.

Jenner grew up in a village in the lovely western English hills on the Severn and Avon rivers. He was the youngest son of the rector of a little church in Berkeley. His parents died when Jenner was five, and his large, loving family quickly substituted for the deceased parents, particularly his brother Stephen, who followed his father's path to the Berkeley pulpit. Stephen took the boy in.

Stephen Jenner was in almost every way the ideal guardian for Edward. He was the consummate Christian gentleman: kind, loving, attentive. He was well-read and liberal. It was he who sent young Edward to the inoculation barn to protect his life. It was he who encouraged Edward's love of nature and curiosity about natural history in an age when science was burgeoning and superstition was in slow, reluctant retreat.

Historian Will Durant pictured it as an age when,

> Two priesthoods came into conflict: the one devoted to the molding of character through religion, the other to the education of the intellect through science. The first priesthood predominates in ages of poverty or disaster when men are grateful for spiritual comfort and moral order; the second in ages of progressive wealth, when men incline to limit their hopes to the earth.

This conflict was easily resolved in Stephen Jenner and the resolution was passed on to his younger brother. When it was clear that Edward had no religious vocation (and probably wouldn't have been able to make a living) Stephen sent him to study medicine with a friend, Daniel Ludlow, in Sudbury.

Ludlow was a surgeon, which did not mean what it does today. A surgeon then was not a medical doctor in the sense that he had a degree from a medical school. A surgeon had learned his medicine as apprentice to another surgeon.

Ludlow was, by all accounts, a good surgeon, and young Jenner learned much from him. It was conventional medicine that Ludlow practiced, which meant that there were particular treatments for each disease and if the treatment failed the fault lay with the patient's constitution, not the surgeon's ministrations. Ludlow was perhaps unusual in that he had a strong sense of cleanliness which Jenner picked up from him.

Ludlow's practice was large, and the pair ranged over Gloucestershire. It was in this practice, probably, that Jenner heard accounts that cowpox, a disease that struck both cows and people, made humans immune from smallpox for the rest of their lives. Ludlow, however, assured Jenner that he knew of no evidence to support the notion. Besides, like most physicians, Ludlow said that he had encountered several cases of people who had smallpox after cowpox, so the assertion could not be true.

Nonetheless the tales prompted Jenner's curiosity. He filed the information for the future.

When Edward had learned all he could locally, Stephen Jenner and Ludlow arranged for him to go to London and study with John Hunter, one of the famed Hunter brothers who were busy revolutionizing anatomy.

After the cleanliness and serenity of Gloucestershire, young Jenner found London a constant irritation. But the man he came to study with was one of the great characters in the history of English medicine, and the apprenticeship began one of the most fascinating scientific partnerships ever. It ended only in 1793, when Hunter died while throwing a tantrum after someone at a medical meeting had contradicted him.

Hunter's brother William ran a world-renowned anatomy institute that was also available to Jenner. Studying anatomy was difficult at the time, because society frowned upon the use of cadavers. Most of the bodies were obtained from the gallows. It was customary for agents of the medical schools to attend all the hangings. While the criminal was still twitching and jerking on the noose, the agents began negotiating with his family to obtain the corpse. The going price was usually about five pounds. Most of the medical students of the day learned their anatomy on bodies with rope burns and broken necks.

It was important to Jenner's development as a scientist that John Hunter practiced unconventional medicine for his time. If a treatment failed, Hunter believed, the trouble was with the treatment, not the patient.

Jenner moved into Hunter's house on Jermyn Street. He never learned to love the city, saying once: "When I write in London my brain seems full of smoke which is too apt to cloud our best faculties."

He and Hunter instantly became friends. They shared both a love of medicine and a passion for natural history. Hunter kept a small museum of fossils and stuffed fauna at his house, which was filled with only a fraction of his collection. He let Jenner work on the collection in his spare time. Later he recommended the young man to Sir Joseph Banks as an excellent curator of the unassorted collection brought back to England from the South Seas by Captain James Cook in the *Endeavour.* Jenner did so well that Banks and Cook offered him a berth on the next Cook expedition, but Jenner turned them down. He still preferred medicine to natural history.

Hunter himself was an irascible, ill-tempered, obstinate man, but he was impressed with Jenner's open mind.

Two years after his apprenticeship had begun, Jenner turned down an offer from Hunter to join him as an associate because he wanted to return to his beloved Berkeley, and in 1773 he set himself up in his brother's house as the town's only surgeon.

He was something of a minor rake at first, riding with the young Earl of Berkeley, but generally he settled in to the quiet existence of a country doctor. He quickly won the respect of other members of his profession and his patients, and kept up his correspondence with Hunter.

It was an amazing correspondence and relationship. The two men, products of their inquiring age, set about to perform dozens of experiments in natural science to learn about the migratory habits of freemartins, the hibernation of hedgehogs, and similar phenomena. Jenner frequently sent specimens to London, including live animals and the skeletons of several porpoises packed in barrels.

After Jenner returned to Gloucestershire, he and Hunt corresponded and occasionally met. One meeting, in 1778, was particularly sad for Jenner. Hunter was obviously suffering from angina pectoris, and although Jenner did not understand the cause of the pain, he knew enough to recognize it as a probably fatal symptom. He brooded for a while and then decided not to say anything to Hunter or his wife lest he upset the old man.

A few years later, while dissecting hearts from patients who had suffered from angina, Jenner discovered the cause: hardening and blockage of the coronary arteries, which diminished the blood supply to the heart and could lead to heart attacks. Hunter was still alive, and Jenner refused to publish his findings so as not to disturb Hunter. It was not important to him that he had made a major medical discovery.

Jenner's association with Hunter eventually led to his becoming a fellow of the Royal Society. The work that earned this honor was a study of the cuckoo, now a classic in zoological history, published in the society's journal.

Naturalists knew that the cuckoo never makes its own nest but deposits a solitary egg in the nests of other birds. The nesting birds' own eggs or hatchlings would disappear, leaving only the young cuckoo to be cared for by its foster parents. Hunter charged Jenner with solving the mystery. Jenner–with his nephew Henry–determined that shortly after hatching the baby cuckoo throws its nestmates or the unhatched eggs out of the nest.

Jenner, who had acquired a formal medical degree by mail–which was standard practice–already could use M.D.; now he enjoyed the honor besides of having the Royal Society's F.R.S. after his name.

At the age of forty Jenner married Katherine Kingscote, a beautiful, gentle woman of some property. They were happily married.

Years into his practice, Jenner still remembered the folk wisdom about cowpox. He practiced variolation but without enthusiasm. If the folk belief was true, here was certainly a better way of protecting people from smallpox. His interest was rekindled by an epidemic of smallpox in 1778, which he believed offered some empiric evidence.

> My attention to this singular disease was first excited by observing, that among those whom in the country I was frequently called upon to inoculate, many resisted every effort to give them smallpox. These patients I found had undergone a disease they called cowpox, contracted by milking cows affected with a peculiar eruption on their teats.
>
> On enquiry it appeared that it [cowpox] had been known among the dairies from time immemorial, and

> that a vague opinion prevailed that it was preventive of the smallpox. This opinion I found was comparatively new amongst them, for all the older farmers declared they had no such ideas in their early days. . . .

He thought that the reason for the recency of this belief was that until the Suttons made variolation relatively safe, nobody had considered immunity to smallpox possible.

The problem with his cowpox theory, Jenner found, was that there was a very large number of exceptions that had to be explained. ". . . I found that some of those *who seemed to have undergone the cow-pox,* nevertheless, on inoculation with the smallpox, felt its influence just the same as if no disease had been communicated to them by the cow."

The consensus of the doctors in the country around him was that Jenner had lost his senses. With few exceptions they scoffed at the contention and resented the man as something of a bore, pestering them with questions. But he persisted. Why he did so in the face of considerable evidence that he may have been wrong is not known. Perhaps it was those weeks in the inoculation barn. Perhaps it was something about scientific method that he had learned from Hunter.

He began to haunt dairy farms looking for cowpox. Some farmers welcomed him. Many thought that he was crazy and chased him off their land. He had determined that he never could penetrate the mystery until he had first studied the cow disease. He continued on and off for years learning that it was really several diseases, not just one. Any outbreak on the udders was called cowpox by the farmers. His research was slowed by the fact that the disease ran in epidemics and several years could go by before cowpox would return to the Gloucestershire farms.

In the summer of 1778 Jenner had been called to the farm

of William Smith. Smith had treated a horse suffering from a disease known as "grease" or "greasy heels" and then had milked the cows. His cows came down with cowpox and so did Smith. Jenner thought that he saw a connection and deduced that grease, cowpox and smallpox were all the same disease. He felt that perhaps it was manifested differently in different animals. (He happened to be wrong, and this error not only slowed his work but made him vulnerable when he finally published his findings.)

He found another case of cowpox the next year, which reinforced his theory. By the time the smallpox epidemic was over in 1782 he had enough information on cowpox to discern the true disease from what he called the "spurious" kind. He had an artist, Cuff, draw pictures of the udders of diseased cows to demonstrate the differences. He never ceased badgering his colleagues, who threatened to throw him out of their medical society if he didn't drop the subject. They all had seen too many examples of people who had cowpox coming down with smallpox and did not perceive a difference between types of cowpox. None had taken the time to examine the cows.

The cowpox epidemic also subsided, leaving Jenner despondent. But Katherine encouraged him to continue his work. When a son was born Jenner inoculated him with grease but nothing was proven by the experiment.

There were all those exceptions to the folk belief to consider. Surely he had missed something.

Jenner resumed his work in 1791, when another cowpox epidemic hit the farms. But it was clear to him by then that there were too many cases of "true" cowpox that failed to give immunity to smallpox.

The year 1793 was a sad one. Hunter died, and Jenner mourned his old friend for months.

But one night in that year, while Jenner was working late

in the barn with some infected cows, the possible solution hit him.

He decided that it all depended on when one contracted cowpox. The disease-laden matter in the pustules varied in strength. If it was administered a day earlier or later than when at its maximum strength, it would be too weak to protect from smallpox.

> During the investigation of the casual cowpox I was struck by the idea that it might be practicable to propagate the disease by inoculation, after the manner of the smallpox, first from the cow; and finally from one human being to another. I anxiously waited for some time for an opportunity of putting this theory to the test.

In May of 1796 a milkmaid named Sarah Nelmes contracted cowpox. She had cut her finger on a thorn before milking infected cows and soon "a large pustulous sore and the usual symptoms accompanying the disease were produced in consequence." Two smaller sores appeared on her wrist. The disease was nearing its zenith. He asked her to come to his office in a few days, when he estimated the sores would be at their worst.

He then sought out eight-year-old James Phipps, a local boy, and obtained his parents' permission for the experiment. Phipps had never suffered from either cowpox or smallpox.

When the milkmaid arrived at Jenner's surgery on May 14, 1796, Jenner took pus from the sore on her hand. He then made two very small scratches on the boy's arm, barely breaking the skin. Jenner let the boy go home and checked him every day.

> On the seventh day he complained of uneasiness in the axilla [armpit] and on the ninth he became a little chilly, lost his appetite, and had a headache. During the whole of this day he was perceptibly indisposed, and spent the night with some degree of restlessness, but on the day following he was perfectly well.
>
> The appearance of the incisions in their progress to a state of maturation were much the same as when produced in a similar manner by variolous matter.

Phipps recovered as expected, and on July 1 Jenner repeated the procedure on him, this time using smallpox matter.

> Several slight punctures and incisions were made on both his arms, and the matter was carefully inserted, but no disease followed. The same appearances were observable on the arms as we commonly see when a patient has had variolous matter applied after either the Cow Pox or the Small Pox. Several months afterwards he was again inoculated with the variolous matter, but no sensible effect was produced on the constitution.

Phipps lived to a ripe old age and was variolated about twenty times to demonstrate his immunity. No variolation took.

The important thing was that Phipps was now immune to smallpox without actually having had the disease! Jenner reported his experiments to a friend on July 19 and pledged: "I shall now pursue my experiments with redoubled ardour."

He could not pursue his experiments for another two years, however—years that must have been agonizing for him—because cowpox did not return until then. But on March 16, 1798, he inoculated two children. One was given matter from the hand of a man who had been infected with grease, the other with pus taken directly from an infected cow. The first boy died of a probably unrelated fever. The second, William Summers, came down with cowpox.

Jenner took pus from Summers and passed it on to a man, William Pead. From Pead the disease was passed on to several others, including seven-year-old Hannah Excell, who, in turn, provided infectious material for four people, including Jenner's youngest son, Robert. Robert's inoculation did not take but the other three came down with cowpox. Jenner did not pursue the matter further, which greatly weakened his case in the coming controversy.

> After the many fruitless attempts to give the smallpox to those who had the cowpox it did not appear necessary, nor was it convenient to me to inoculate the whole of those who had been the subject of these late trials, though I thought it right to see the effects of variolous matter on some of them, particularly William Summers, the first of these patients who had been infected with matter taken from the cow. He was therefore inoculated with variolous matter from a fresh pustule but as in preceding cases the system did not feel the effects in the smallest degree.

That ended the problem as far as Jenner was concerned.

> While the vaccine discovery was progressive, the joy I felt at the prospect before me of being the instrument destined to take away from the world one of its greatest calamities, blended with the fond hope of enjoying independence and domestic peace and happiness, was so

> excessive that, in pursuing my favourite subject among the meadows I have sometimes found myself in a kind of reverie. It is pleasant for me to recollect that these reflections always ended in devout acknowledgments to that Being from whom this and all other miracles flow.

Jenner's next problem was how to tell the world. He took some of his notes and showed them to friends, who suggested that he try to get them published by the Royal Society. But the society found the concept too revolutionary and the supporting data a bit weak and, to Jenner's surprise and disgust, rejected his paper.

His friends then advised him to get it published on his own. Jenner rode into London to the firm of Sampson Low on Berwick Street and ordered his paper printed. It appeared in June of 1798. It is one of the most famous works in medicine: *An Inquiry into the Causes and Effects of the Variolae Vaccinae; A Disease Discovered in Some of the Western Counties of England, Particularly Gloucestershire, and Known by the Name of the Cow Pox,* By Edward Jenner, M.D. F. R. S. &C.

The title page contained a quotation in Latin from Lucretius which suggests that in distinguishing between truth and falsehood man must use his senses.

It was a very small book, sixty-four pages of large type in the first edition, and an immediate best-seller, finding its way into the salons and drawing rooms of the elite of London and into the medical establishment and the ladies and gentlemen of the Enlightenment. How many actually read it is unknown, but it made Jenner instantly famous and vaccination (a term he coined) one of the great controversies of the time. It begins:

> The deviation of man from the state in which he was originally placed by nature, seems to have proved to

> him a prolific source of diseases. From the love of splendour, from the indulgence of luxury, and from his fondness for amusement, he has familiarised himself with a great number of animals which may not originally have been intended for his associates. The wolf, disarmed of ferocity, is now pillowed, in the lady's lap. The cat, the little Tyger of our land, whose natural home is the forest, is equally domesticated and caressed. The Cow, the Hog, the Sheep, and the Horse, are all, for a variety of purposes brought under his care and domination.

Jenner, reflecting the theology of the day, is saying that diseases are caused by man's lust for civilization and ease. Having made his bow to original sin, he then lays out his theory that smallpox, cowpox and grease are somehow related. He links that chain to immunity. He then draws the distinction between "spurious" cowpox and the real thing. He finally gets to the cases, listing fifteen people who acquired cowpox and subsequently proved to be immune from smallpox when variolated. The last person, Thomas Pearce, the son of a blacksmith who caught grease from a horse, posed something of a problem for Jenner: "It is a remarkable fact, and well known to many, that we are frequently foiled in our endeavours to communicate the smallpox by inoculation [variolation] to blacksmiths . . . yet the following case renders it probable that this cannot be entirely relied upon. . . ." Jenner admitted that Pearce subsequently came down with smallpox, but it did not shake his confidence in his theory.

Jenner reiterated the theory which blamed the horse grease and suggested (incorrectly) that many common diseases, including measles and scarlet fever, also have a similar cause. He then explains why his procedure improves on variolation:

> . . . notwithstanding the happy effects of inoculation, with all the improvements which the practice has received since its first introduction into this country it not very unfrequently produces deformity of the skin, and sometimes, under the best management, proves fatal . . . but as I have never known fatal effects arise from the Cow Pox even when impressed in the most unfavourable manner, producing extensive inflammations and suppurations on the hands; and as it clearly appears that this disease leaves the constitution in a state of perfect security from the infection of Small Pox may we not infer that a mode of inoculation may be introduced preferable to that at present adopted, especially amongst those families which from previous circumstances, we may judge to be predisposed to have the disease unfavourably?
>
> [Jenner lists several other reasons why his method is better than variolation and concludes]: I shall myself continue to prosecute this inquiry, encouraged by the pleasing hope of its becoming essentially beneficial to mankind.

Modern scientists can find much to criticize in the *Inquiry,* just as some of Jenner's contemporaries did. For one thing, Jenner was not really the first man to inoculate with cowpox to protect against smallpox. In 1774 a cattle dealer, Benjamin Jesty, inoculated his wife and three sons with cowpox. Jesty, however, was just a cattle dealer and made no attempt to publish what he did or to spread the word in any way. He surfaced later in the ensuing battle over vaccination. Jenner, however, was the first scientist to come up with the notion, study it by systematic experimentation, and bring it to the attention of his contemporaries.

Another objection is that the *Inquiry* discusses only one

case—that of James Phipps—in which cowpox was deliberately induced from another human and variolation then attempted. In every other case Jenner either merely administered cowpox to patients or tried to vaccinate people who had acquired cowpox by ordinary infection from humans or from cows. After Phipps' case Jenner didn't think that it was necessary to prove his point. This failure to produce more than one case weighed heavily in the decision by the Royal Society not to publish his paper.

Jenner also kept insisting on linking cowpox and smallpox to grease. He lists three men who contracted grease in the *Inquiry:* one subsequently caught smallpox, one was successfully variolated and one seemed to be immune. Jenner said that this failure resulted because the agent causing the three diseases must be passed through a cow before it takes on its immunizing properties—an ingenious rationalization.

Jenner also failed to mention the death of the boy who had grease, a serious omission even if he felt that it was unrelated. He merely wrote that "the little boy was rendered unfit for inoculation."

All of the above notwithstanding, Jenner was right, even if for the wrong reasons. Vaccination with cowpox virus does confer immunity to smallpox and does so safely and easily and with almost 100 percent effectiveness. It proved to be just the godsend Jenner hoped for and provided the principal weapon for the eventual death of the disease.

In 1800 Jenner published a second edition of the *Inquiry* adding several observations on "spurious" cowpox and responding to a letter from James Earle of Frampton-upon-Severn, Gloucestershire, who reported to him a death from vaccination. Jenner blamed imprecise procedures:

> Certain . . . is it that variolous matter may undergo such a change from the putrefactive process, as well as from some of the more obscure and latent process of

> nature, as will render it incapable of giving cow pox in such a manner as to secure the human constitution from future infection, although we see at the same time it is capable of exciting a disease which bears so strong a resemblance to it, as to produce inflammation and matter in the incised skin (frequently indeed more violent than when it produces its effects perfectly) . . .

But if Jenner thought that his findings would end the problem of smallpox quickly, or be accepted by his peers or the public without a bruising fight, he was sadly mistaken. In many ways his work had just begun. But the weapon for the defeat of smallpox had been found.

VII

The Bells of Boston

On April 22, 1721, His Majesty's Ship *Seahorse,* under the command of Captain Wentworth Paxson, turned the corner around Cape Cod and sailed into the harbor of Boston, Massachusetts Bay Colony. No illnesses were reported, and the ship was permitted to anchor without quarantine.

But before it could sail again a black seaman came down with smallpox. He was quickly placed in isolation in a house flying a bright red flag bearing the legend "God have mercy on this house." A special nurse was locked inside with him. Two guards were posted outside.

Several days later another black came down with smallpox and also was quarantined. Soon the sick list was up to eight, where it stabilized for some time. It had been twenty years since the last outbreak in Boston and the report of eight new victims did not appear to frighten the town.

In May the epidemic exploded. By midsummer businesses

were closed and all public assemblies but church services were banned. Only survivors of earlier epidemics, particularly the last one in 1702, were safe. Everyone under nineteen was at risk.

Church bells tolled death so often that summer that the selectmen ordered that they be rung only once for each smallpox victim, and then only at designated times. But there were so many deaths, and the funeral habits were so ingrained, that the rule was universally disobeyed and the sorrowing bells of Boston rang day and night.

About a tenth of the population fled the city. Ships and merchants refused to enter Boston and special relays had to be set up so that needed commodities such as food and firewood could be brought in. Tradesman would bring their goods to an intersection and walk away. Men from Boston could then come and pick them up. Similar relays using small boats were set up in the harbor.

The ten or so physicians in the city, totally unable at first to help in any meaningful way, went from house to house trying simply to comfort the sick.

The epidemic of 1722 became one of the worst disasters in Boston's history. The city capitulated to hysteria, fueled by an already-century-old history of smallpox terror in America.

Smallpox had arrived in the British colonies with the very first whites. Two years before the Plymouth landing an English adventurer, Captain Dermer, passed the winter of 1618-1619 in a New England harbor and watched smallpox wipe out a large portion of the Indian population. (His journals call the disease "plague," but his description is of smallpox.) When the Pilgrims arrived they would note the effects of this depopulation. The tribes had been reduced from about nine thousand souls to a few hundred. The Massachusetts tribe was believed to have lost twenty-seven hundred out of three thousand members.

Smallpox was with John Winthrop's seventeen-ship fleet of 1630. One passenger, Francis Higginson, wrote that on his ship several people were stricken, "yet thanks be to God none dyed of it but my owne childe." Another Puritan on another vessel reported fourteen deaths on the voyage and said that "we ware wonderfule seick as we cam at sea withe the small Poxe."

It is possible that many of the settlers had acquired some immunity during their transatlantic crossings. Nonetheless, smallpox became one of the hazards the early colonists learned to expect from life, along with blizzards, Indian attacks, famine, floods and fires. For the Indians, the disease often amounted to a catastrophe. William Bradford described one scene near the Plymouth Plantation:

> They [the Indians] fell down so generally of this disease as they were in the end not able to help one another, nor not to make a fire nor to fetch a little water to drink, nor any to bury the dead; but would strive as long as they could, and when they could procure no other means to make fire, they would burn the wooden trays and dishes they ate their meat in, and their very bows and arrows. And some would crawl out on all fours to get a little water, and sometimes die by the way and not to be able to get in again.

There were outbreaks in Massachusetts again in 1638 and in 1639. In 1648 there was a general epidemic, but the disease was not reported to be very virulent in most places, one exception being Cape Cod, which declared a "Day of Humiliation" on November 15. Children on the Cape seemed to suffer the most because the disease was accompanied by whooping cough.

For another twenty years, the larger towns of the Bay Colony were free of smallpox. But in the woods the French and Indians usually had the disease. In 1666 one of them—or perhaps a shipload of immigrants—carried the disease back into Boston.

This outbreak was fairly severe, with hundreds of cases but only forty deaths, according to one report. There was another mild epidemic in 1675, probably contained by quarantine procedures. But the young Cotton Mather, who would play such a vital role in the 1721 disaster, warned from his pulpit that "very soon God will lift up his hand against Boston." Two years later smallpox struck again.

The disease came by ship, with Charlestown suffering first in the winter and Boston by the spring. By June 6 there had been eighty deaths in the two towns and a fast was proclaimed. According to Increase Mather (Cotton's father), the deaths numbered seven or eight a week in Boston during the summer, and in September it grew much worse. On September 30, thirty people died. In a four-week period a hundred fifty deaths were officially recorded, but one witness said that there were twice that number. Travel was restricted and towns outside the Bay area quarantined themselves against travelers from Boston.

This epidemic led to the publication, in 1677, of the first medical text in America, Thomas Thacher's *A Brief Rule to Guide the Common People of New England How to Order Themselves and Theirs in the Small Pocks, or Measles.*

The next epidemic was in 1689, when slaves from the West Indies carried the disease to Boston by ship and it soon spread as far as Canada and New York.

In September 20, 1690, the first American newspaper, *Publick Occurences,* in its first and last edition, reported:

> The smallpox which has been raging in Boston, after a manner very extraordinary, is now very much abated. . . . The time of its being most mortal was in the

> months of June, July, and August when sometimes in the Congregation on Lord's Day there would be Bills desiring prayers for above one hundred sick. It seized upon all sorts of People that came in the way of it, even infecting children in the bowels of their mothers that had themselves undergone the Disease many years ago; some were born full of the Distemper. . . .

That report was probably one of the reasons the newspaper was suppressed by the authorities. The colonists did not like to advertise pestilence. It was bad for business.

Ten years later sailors returning to port in Stonington, Connecticut, brought the disease in with them. The epidemic there was localized by quarantine and only four died.

The last major outbreak in New England before the tragedy of 1721 was in 1702. Cotton Mather's journals are filled with horror as the disease spread. "More than fourscore people were, in this black month of December, carried from this town to their long home . . ." he wrote.

The population of Boston in 1700 was about seven thousand, so that an epidemic in which three hundred died was not unusually large (4 percent or so). It was merely a prelude. The same year one quarter of the population of Quebec was dying off in a possibly related epidemic.

The middle colonies may have had an even worse time with smallpox than New England. The main vector, or path of transmission, may have been the Indians, and the catalyst was the recurring warfare among the French, the Indians and the English and Dutch settlers.

During the period 1688-1691 smallpox seemed almost endemic to most of the middle colonies. With each outbreak of warfare, such as King William's War, the disease spread from army camp to army camp, and then traveled back to the towns when the army dispersed. Frequently it influenced military events. Quebec was saved from invaders by the virus a number of times. On one occasion a colonial general lamented that he had tried to stir up the Indians for yet another invasion of French America but the Indians would have none of it. "The smallpox have carry'd away some & divers of more of them have been . . . sick." His superior suggested that the disease was probably "God's anger."

New York bore the brunt of many of these outbreaks. Colonial troops near Albany were particularly ravaged, and they spread the disease every time they changed camp.

The southern colonies escaped much of the horror of the disease, but even there deaths were not uncommon. By 1696 there had been outbreaks reported in Virginia and South Carolina, with the South Carolina Council writing to the Lord Proprietors in London that "we have had the small pox amongst us nine or ten months, which hath been very infectious and mortal. We have lost by the distemper 200 or 300 persons." Once again the Indians suffered more. One tribe, the Pemlico, disappeared in the epidemic.

In 1711 William Byrd of Virginia reported that there were two particularly tragic cases in that colony. Because no one dared to venture near the victims, they both died of neglect.

That same year South Carolina was hit again. One official wrote: "Never was there a more sickly or fatal season than this for the small pox, pestilential feavers, pleurisies, and fluxs have destroyed great numbers here of all sorts, both whites, blacks and Indians, and these distempers still rage to an uncommon degree."

So the terror of smallpox was hardly foreign to Bostonians by the spring of 1721. Nor were they completely unarmed. From the middle of the seventeenth century there had been a series of public health regulations designed to quarantine carriers of infectious diseases. Since the New England colonies were the best organized of the British outposts, these measures succeeded frequently in preventing epidemics. For instance, in 1648 the General Court of Massachusetts ruled that all ships heading into Boston anchor three miles offshore and allow "no persons or goods from the ship be brought ashore without the permission of three members of the Council." The ships were inspected for disease.

The selectmen in Boston once ordered that bedding not be aired in front yards or on the roads during epidemics. Another rule required that the coach between New York and Boston be fumigated regularly. They were also quick to prohibit entry to people who might be bringing the disease from outside areas—almost as quick as their neighbors were when smallpox broke out in Boston. In 1663, during an outbreak in New York, the Massachusetts Bay General Assembly enacted a law beginning:

> This court, understanding that the hand of God is gone out agaynst the people of New Netherlands, by Pestilential infections, doe therefore prohibit all persons from comeing from any of those infectious places into this Colony, and amongst our people, until ye Assistants are informed and satisfied that the distemper is allayed.

In 1707 a Massachusetts law provided that when persons "were visited with the plague, smallpox, pestilential fever and other contagious sickness" the local selectmen were empowered to quarantine them into separate houses, provide nurses and guards, and charge the victims for the cost.

Other colonies passed quarantine laws also. In 1721 Captain Joseph Allyn landed at Wethersfield on the Connecticut River with smallpox reported on his sloop, and the governor and council of Connecticut colony ordered that:

> . . . The doors and windows of Mr. Allyn's house next the street, and at each end, be nailed up and so effectually secured as to prevent anything being conveyed

> into or put out of the house on the side next the highway, or towards the neighbouring houses at each end; that care be taken to let in sufficient air on the backside of the home.
>
> That the tenders on the sick, or nurses, be charged that whatever they have occasion to bring out of the sick person's room and throw out of doors, be carried out some back way, and in some convenient place for that end buried, or covered with dirt, to prevent the dilating of any scent in the air.

In October of that year, with smallpox raging in Boston, John Rogers of New London returned to Connecticut with smallpox. There were at least twelve meetings of the governor and council devoted almost entirely to the "stupidity and stubbornness" and "unruliness" of the family and friends of the patient who repeatedly broke quarantine. Finally two guards were assigned to move into the neighborhood to keep "watch and ward day and night, and by coming as near to the house of said Rogers as they may without danger of infection, labour to understand the state of the sick there" and to discover and prevent any communications between those in the house and the outside world. Dogs that roamed the neighborhood, including the Rogers' family pets, were ordered destroyed.

The earliest pesthouse to isolate cases was built in Massachusetts on Spectacle Island in 1717. Later it was transferred to Rainsford Island. Other colonies soon followed the lead. These houses had little actual value to the patients, although medical care was no worse than they would have received outside, but the spread of the disease was cut down.

Boston, therefore, was not unprepared for smallpox. Indeed, it had two other weapons of considerable might: the Reverend Cotton Mather and Dr. Zabdiel Boylston.

What these men shared was a devotion to New England and to her people. Even at this early stage of American history—exactly a hundred years had passed since the first group of colonists arrived—it was possible to have deep roots in the rocky, forested soil of Massachusetts. Both men did.

Boylston was born in either 1669 or 1670, the son of Dr. Thomas Boylston, the physician and surgeon, in the wilderness settlement of Muddy River (Brookline). He was the seventh of twelve children. It is believed that he apprenticed with his father until the elder Boylston died when Zabdiel was fifteen. Somewhat later (his early years are not well documented) Zabdiel went to Boston and studied with John Cutler, considered the best surgeon in the area. Cutler, who was Dutch, had settled in Boston in 1694 and set up his practice there. He and Boylston apparently got on well together, because the records show that Boylston remained with Cutler for several years after his training would have ended.

In 1706 Boylston married and set up his practice on Dock Square. He also established a large apothecary shop, inspired by his experience with his father, who had made most of his own remedies from what he could find in the forest.

There were ten or eleven other physicians in Boston in 1721. We know the names of ten. One was William Douglass, a Scotsman and a bit of a snob, who never quite managed to accept his life as a New Englander and had difficulty in taking any of his contemporaries seriously unless they had a university background. He and the field-trained Boylston were predestined to clash.

Others included Thomas Robie of Cambridge, Harvard and the Royal Society; Thomas Bulfinch, an apprentice of Boylston; Oliver Noyes, who was to die during the epidemic; John Perkins, a friend and neighbor of Mather; and doctors named White, Cooke, Oakes and Williams. Probably there were some others. The classification of who was or was not a doctor was extremely fuzzy in colonial America.

Cotton Mather's Massachusetts roots went even deeper than Boylston's. He was the grandson of the Puritan minister Richard Mather, who settled in Massachusetts in 1635 to preach in freedom. Richard Mather's youngest son was Increase Mather, minister, diplomat and president of Harvard. Increase was educated at Trinity College in Dublin, refused a post there, and returned to Boston in 1661 to become pastor of North Church.

Cotton Mather (his first name was his mother's maiden name) was born in Boston on February 12, 1663, and entered Harvard at the age of twelve, receiving his diploma six years later from his father, the president. He once declared that his life was "a continual conversation with heaven."

He and his father came by their unhappy reputation in history partially because of their participation in the witch trials. Their belief in witchcraft was shared by almost everyone who lived then. A book by Increase Mather, *An Essay for the Recording of Illustrious Providences,* published in 1684, probably provided the rationale for the trials that followed. On the other hand, both Cotton and Increase fought against what was called "spectral evidence," which was testimony of an alleged victim who claimed to have been attacked by a specter which looked exactly like the defendant. The change in the rules of evidence pushed by the Mathers made it a bit harder to convict a suspected witch and to fake witchcraft evidence.

Mather was not out of character in entering the scientific dispute that erupted in the middle of the 1721 epidemic. He was well-read and interested in all things scientific. He and his father, moved by a fascination for natural science and God's role in it, had prepared a list of unexplained natural phenomena in America called *Curiosa Americanae,* which had been published in the *Philosophical Transactions* in 1716 and earned Mather fellowship in the Royal Society. Furthermore, at one time Mather had considered a career in medicine, particularly when he was young and his stammer made success in the pulpit unlikely. Eventually he cured himself of the stammer but kept his interest in medicine.

Cotton Mather's involvement in the smallpox controversy actually began on December 15, 1706, when, according to his diary, "Some gentlemen of our Church, understanding . . . that I wanted a good *Servant* at the expense of between forty and fifty pounds, purchased for me a very likely *Slave,* a young man, who is a Negro of a promising aspect and temper, and this day they presented him unto me. I putt upon him the name of Onesimus."

Onesimus probably was fresh from Africa, and Mather needed to know what diseases he might have had or might be susceptible to. When he asked about smallpox he received an unexpected answer. Onesimus had been variolated by his tribe. The practice was not uncommon in Africa.

Fourteen years later Mather read a copy of the *Philosophical Transactions* and came upon the Timonius article. He began a correspondence with Dr. John Woodward of the society in which he recalled Onesimus' response to his question:

> I am willing to confirm you, in a favourable opinion, of Dr. Timonius' Communication; and therefore, I do assure you, that many months before I mett with an Intimation of treating ye Small-Pox with ye Method Of Inoculation, I had from a Servant of my own, an Account of its being practiced in *Africa.* Enquiring of my Negro-Man Onesimus, who is a pretty intelligent fellow, whether he ever had ye Small-Pox, he answered *Yes* and *No;* and then told me that he had undergone an Opera-

> tion, which had given him something of ye Small-Pox, and would forever preserve him from it, adding, That it was often used among ye *Guaramantese,* and whoever had ye Courage to use it, was forever Free from ye Fear of the Contagion. He described ye Operation to me, and showed me in his Arm ye Scar, and his description of it made it the same that afterwards I found related unto you by your Timonius.

Mather had had smallpox when he was fifteen. His family had a number of close brushes with the disease. He was impressed by the evidence that variolation worked, and he ended his letter to Woodward with these words: "For my own part, if I should live to see ye *Small-Pox* again enter o'r City, I should immediately procure a consult of o'r physicians to introduce a practice, which may be of so very happy a tendency."

He got his chance five years later.

At the time variolation was apparently almost entirely unknown in the Americas. The debate was just breaking out in England when the *Seahorse* docked in Boston harbor. When the disease erupted, however, Mather remembered his promise. He noted in his diary on May 26, 1721: "I will procure a consult of our physicians, and lay the matters before them." On June 6 he wrote a letter to members of the Boston medical community, which was passed from one to the other.

Their response must have shocked him.

The life of the city already had been completely disrupted by the epidemic. At least half the residents had fled, or were about to. The city Mather deeply loved was dying. In his letter he did not urge the physicians to use inoculation; he merely wanted them to get together and consider it. He even did some homework for them, borrowing a copy of the *Transactions* from Douglass.

But no one came to his "consult." Only Dr. White responded with any interest. Whether the physicians honestly rejected inoculation as a public health measure, or resented interference from an outsider, is not known. The only action his letter produced was a demand from Douglass that Mather return the copy of the journal. Douglass subsequently refused to show the publication to anyone, even the governor of Massachusetts, who asked to see it when the controversy exploded in the city.

Mather proceeded to inform his friend, Zabdiel Boylston, about variolation:

> You are many ways, Sir, endeared unto me, but in nothing more than in the very much good which a gratious God employes you and honours you in a miserable world. I design it, as a testimony and esteem that I now lay before you, the most that I know (and all that was ever published in the world) concerning a matter, which I have the occasion of its being much talked about. If upon mature deliberation, you should think it admissible to procede in, it may save many lives that we set a great store on. But if it not be approved of, you will have the pleasure of knowing what is done in other places.

"Upon reading [the letter from Mather]," Boylston wrote, "I was very well pleas'd and resolv'd in my mind to try the experiment." Soon after, early in July, Boylston made the trial.

Boylston had contracted smallpox as a youth, so he was already immune. But his six-year-old son, Thomas, was not. Nor were his slave, Jack, thirty-six, or Jack's son, Jacky, two and a half. He inoculated them with scabs from a patient. By sunset the entire city knew what he had done and the immediate reaction was shock. That Boylston would endanger the life of his own son with an unproven technique opposed solidly by the medical community was appalling to many Bostonians. Wasn't he aware of the danger to the boy and to the city if the boy's disease should start a whole new chain of infection?

Mather was surprised at the outrage. "They rave, they rail, they blaspheme; they talk not only like ideots, but also like *fanaticks,* and not only at the physician, who begun the experiment, but I also am the subject of their fury; their furious obliquies and invectives."

The violence of the public reaction served as a counterpoint to the terror that stalked the city. On July 13 a day of humiliation was ordered through Massachusetts, with city-wide religious services. The number of cases was mounting steadily and the city was paralyzed.

The clamor over inoculation increased with the epidemic. Finally it became impossible for the officials to ignore the dispute and a public meeting was ordered for the 21st. The medical community was invited to give testimony. Public officials and private citizens also were invited. The main testimony was a report translated from the French written by a doctor named Lawrence Dalhonde, on the long-range effects of variolation that he claimed to have seen in the French army. Dalhonde asserted that many soldiers who had been inoculated eventually died of sundry causes. Boylston later described the report as unbelievable.

But the selectmen of Boston did not see it that way. They believed what the bulk of medical testimony assured them was true, and passed the following resolution on July 22:

> A resolve upon a Debate held by the Physicians of Boston concerning Inoculating the Small Pox, on the twenty-first day of July, 1721. It appeared by numerous Instances, That it has prov'd the Death of many Per-

sons soon after the Operation, and brought Distempers upon many others, which have in the End prov'd fatal to 'em.

That the natural tendency of infusing such malignant filth in the Mass of Blood, is to corrupt and putrify it, and if there be not a sufficient Discharge of that Malignity by the Place of Incision or elsewhere, it lays a foundation for many dangerous diseases.

That the Operation tends to spread and continue the Infection in a Place longer than it might otherwise be.

That the continuing the Operation among us is likely to prove of most dangerous Consequence.

Boylston's quiet contribution to the meeting was to inform the selectmen that now he had seven patients under treatment and to invite everyone to visit them to see how well they were doing. No one accepted the invitation. Later Boylston wrote: "It is a thousand pities our select-Men made so slight and trifling a representation of the small pox, that has always prov'd so fatal in New England, as they seem to have done. . . ."

The tumult grew in intensity. Boston, Mather wrote later, had become a "dismal Picture and Emblem of Hell; Fire and Darkness filling of it, and a Lying Spirit reigning there." All the while, he wrote, "widows multiply."

Leading the opposition was Douglass. In a letter to Dr. Cadwallader Colden a few years later, he explained his motive:

I oppose this novel and dubious practice not being sufficiently assured of its safety and consequences; in short I reckon it a sin to propagate infection by this means and bring on my neighbor a distemper which might prove fatal and which perhaps he might escape

> (as many have done) in the ordinary way, and which he might certainly secure himself by removing in this country where it prevails seldom.

That was not irrational considering the somewhat checkered history of variolation. What was irrational was the *ad hominem* assault Douglass launched against Boylston in the press.

From behind the thin veil of a *nom de plume,* he wrote a scathing letter to the Boston *News Letter*, charging that Boylston was ignorant, was mishandling his patients and should be tried for a felony if anyone died as a result of inoculation. (Several would die.) The proposal was so extreme that the debate was stunned into silence for five days. Then six clergymen, including the two Mathers, wrote to the newspaper to defend Boylston:

> It was grief to us the Subscribers, among others of your friends in the Town, to see Dr. Boylston treated so unhandsomely in the Letter directed to you last week and published in your Paper. He is the Son of The Town whom Heaven (as we all know) has adorned with some very peculiar Gifts for Service to his Country, and hath signally own'd in the Successes he has had.

This defense by the clergy highlighted a major difference between the debate in America and that going on in England. In the colonies the clergy strongly supported inoculation, urging it on their parishioners.

Boylston remained the gentleman:

> It might be easy for me to make answers to the Scurrilous things lately published against me, and satisfy the Publick of the Falsehood and Baseness in them, but I think it better becomes a considerate man to

> decline foolish Contentions particularly at a time when there is a *grievous Calamity* upon us, that calls us (instead of railing at one another) to unite in prayers to God for His Mercies to us. And therefore if any think to go on with their Calumnies and Fooleries I shall not think to take any notice of them. What I do (I hope as it has hitherto done) will vindicate itself with People of Thought and Probity.

Mather and Boylston would collaborate on a treatise on variolation before the year was out. Both men's stake in the epidemic was growing.

During the height of the dispute Mather's son, Samuel, returned from Harvard in a state of complete terror. His roommate had just died of smallpox. Samuel wanted his father to have him inoculated. Mather was torn between the principle he had so loudly espoused in public and concern for his son.

"How can I answer it?" he wrote. "If on the other side, our people who have Satan filling their hearts and their tongues, will go on with infinite prejudices against me and my ministry. If I suffer this operation on my Child, and be sure, if he should happen to miscarry under it, my condition would be insupportable."

At the urging of Increase, now approaching ninety and watching the debate, the operation was performed in private. If the boy died Mather could keep it a secret. Samuel had a difficult time but survived.

The bells of Boston were increasing their sad tolling. The only vehicles on the streets now were the death carts. Among the dead were six people who had been variolated. Boylston still stoutly defended the practice:

> As to those who died under Inoculation, I would observe that Mrs. Doxwell, we have great reason to

> believe was infected before. Mr. White thro' splenetic delusions, died rather from abstinence than from Small Pox. Mrs. Scarborough and the Indian girl died of accidents by taking cold. Mrs. Wells and Searle were persons worn out by Age and Disease, and very likely these two were infected before.

Boylston was sure that if he inoculated people who were incubating the disease, the inoculation would not work. He felt that there was a difference in how the body reacted to the variolation and how it reacted to whatever caused smallpox.

"It is my opinion, from reason, and as far as I can Judge, from experience too, that a person may be infected in the natural way," he wrote, "and yet go the usual time of nine days before eruption. And it is further my opinion, that a person may be served by inoculation, if not infected above a day or two before inoculated."

His method was to take "pus from the ripe pustules" and then "take a fine cut, sharp toothpick (which will not put the person in any fear, as a lancet will do many), and open the pock on one side" and scoop the pus into the wound with a quill.

In early October a census was begun by the city fathers to see just how many cases of smallpox there were in Boston. The count soon was hampered by the fact that the surveyors were coming down with smallpox.

The violent outrage against inoculation reached its apogee on November 13—a bomb was thrown through Mather's window at 3 a.m. Attached to the bomb, which did no damage, was a crude note reading: "COTTON MATHER, You Dog, Dam You. I'll inoculate you with this, with a Pox to you."

The governor offered a reward for the capture of the culprit, who was never found, but Mather was delighted. Being a martyr appealed to him enormously. He was already under attack in the press by James Franklin's *New England Courant* (James was Benjamin's brother). But Mather kept turning out broadsides supporting inoculation and encouraging people to take it. His opponents continued their attack, even dragging the rest of the clergy into it. The whole argument began to take on a decidedly anticlerical aspect which annoyed the Mathers.

But by later autumn the debate over variolation abated somewhat as more and more people turned to it. The epidemic also began to slacken as it ran out of potential victims, and some of the stiff quarantine regulations were eased. By May of 1722 the epidemic had ended.

The census, when completed, showed that more than half of the city had been stricken with smallpox. Out of a population of 10,670 there were 5,980 cases (56 percent) and 844 deaths (14 percent). One journalist suggested that those numbers were too low because so many people had fled the city. In fact, only seven hundred of the people who remained in Boston throughout the disaster stayed free of smallpox. Most probably were survivors of the 1702 outbreak.

Those who fled the city, of course, spread the disease all over New England despite the quarantine measures.

It was, however, the last epidemic in the United States in which smallpox was allowed to run unfettered. Every later outbreak, including one nine years later, was modified, at least to some extent, by inoculation. The rate of sick and dead would never be quite as high again.

In 1730 smallpox returned to Boston by ship from Ireland. As before, the press of the city tried to ignore the epidemic until it had spread so far and had become so noticeable that the newspapers' integrity would have been damaged by further suppression. In January of that year two Boston newspapers denied that there were any cases of smallpox in the city. Two months later the entire government of Massachusetts had to flee. Weekly burials averaged about forty. There were about four thousand cases and five hundred fatalities, a death rate slightly below that of 1721. Whether variolation played a role in the reduction is unknown but certainly it was practiced. During the year the Boston selectmen had to pass the following ordinance:

> We being inform'd That many Persons belonging to the Adjacent Towns intend to come into this Town to have the Small Pox Inoculated upon them, which we apprehend would be much to the damage of this place, This is therefore to give Publick Notice, That if any Person or Persons, not belonging to this Town shall presume to come into it upon the aforesaid Occasion, he or they shall be prosecuted according to Law.

In other words, people from the suburbs were now flocking to Boston, or at least threatening to do so, to be inoculated.

Unfortunately Cotton Mather was not there to be vindicated. He died in 1728. Boylston was still alive, however, and might have been pleased by the fact that one of those performing variolation was William Douglass, now a convert to the cause.

Because of the extensiveness of the 1721 epidemic it is likely that most of the victims of the 1730 outbreak were children under the age of nine, the only ones left without immunity. The disease was gone in a year.

Boston was not stricken again until 1751, the disease coming by sea from London. Again there was a concerted effort to deny what was happening, a device that permitted Boston to continue its commerce with other towns for as long as possible. One minister reported, however, that "the smallpox is broke out in Boston in a violent manner after two and twenty years of absence, it has occasioned almost a total stagnation of trade so that half the houses and shops in town are shut up and the people retired to the country."

One place they could not retire to was Newport, Rhode Island, which increased the fine for anyone from Boston trying to get into the city or for those harboring Bostonians. The courts of Boston were shut down and the judges escaped to the suburbs.

The statistics for this episode are quite detailed, thanks to the Reverend Thomas Prince of Boston. Prince recorded that out of a population of 15,684, there were 5,545 cases (35 percent) and 2,124 cases resulting from inoculation. In all 569 people died (7 percent). According to Prince, many people fled the city, and of those who remained and had not been previously infected only 174 escaped.

The odds of survival were definitely better for the inoculated. According to municipal records (which had a different count from Prince's), the death rate for those naturally infected was 9 percent. The rate for those inoculated was only 1.5 percent.

The rise in the use of variolation is shown in the figures of the Massachusetts Historical Society.

DATE	NO. INOCULATED	DEATHS	DEATH RATE
1721	247	6	2.4%
1730	400	12	3
1752	2,109	31	1.5
1764	4,977	46	0.9
1776	4,988	28	0.6
1778	2,121	19	0.9 *

* The decrease in the number of persons inoculated in 1778 was probably due to the large number who were immune by then.

In 1792 more than nine thousand were inoculated and the death rate jumped to 1.8 percent. Historians believe that the increase occurred because the physicians were not as careful as they had been in their procedures and because many people tried variolation after they came down with smallpox, hoping it would lessen their suffering.

Also there is no doubt that variolation helped spread the disease because those inoculated stayed on their feet until they were finally driven to bed by fever. Even then they might sit up in bed entertaining friends. It was the equivalent of several hundred additional cases of smallpox walking around, cases the victims took much less seriously than they did natural smallpox. Eventually governments had to put some restrictions on the practice of inoculation.

Inoculation had also picked up some vital support among the colonies' power structures by the second half of the century. In 1730 Benjamin Franklin reported in the *Pennsylvania Gazette* in Philadelphia that the "practice of inoculation for the smallpox begins to grow among us. J. Crosden, Esquire, the first patient of note, is now upon recovery, having had none but the most favorable symptoms during the whole course of the distemper, which is mentioned to show how groundless all those reports are that have been spread through the Province to the contrary." In 1736 one of Franklin's sons died of smallpox and rumors spread that he had been inoculated. Franklin, who supported inoculation, quickly denied this in the *Gazette*. He reported later that of the hundred twenty-six persons inoculated in Philadelphia during an outbreak in 1736-37, only one had died.

Another man who supported variolation was young Thomas Jefferson. In 1766 Jefferson, then twenty-three, went on a leisurely three-month voyage to Annapolis, Philadelphia and New York, in part to find someone to inoculate him.

As it became obvious that inoculation worked its supporters became bolder. In 1756 a Philadelphia surgeon, Lauchlin

Macleane, issued a broadside in favor of variolation. Aside from the fact that doctors care about the health and lives of their patients and their community, he wrote, "there is yet another reason not to doubt the candour of physicians, and that is a strong one, when they speak in praise of inoculation; I mean, that such an opinion is contrary to the interest of their pocket, for before the practice of inoculating was introduced the small pox was certainly the surest and largest penny in the doctor's purse, he being as certainly called for as the disease came. . . ."

> The chief argument used against Inoculation by scrupulous Persons, is from conscience. It is Presumption, say they, to tempt the Almighty by inflicting Distempers without His Permission. So say I, but the great Success of the Practice not only shows the Permission of God for, but his immediate Blessing on our Endeavors by the extraordinary Recovery of so many more, in this, than in the natural way, as it is called, of the Disease.

Similar arguments would be required when Jenner produced his vaccine.

Tragedy and terror did not disappear with the use of variolation. An account of one funeral recorded in the diary of Joshua Hempstead of New London, Connecticut, reported that "Nathanael Coit and the widow Hobbs buryed Hempstead. They drew the coffin with long ropes on the ground" so as not to become infected themselves.

The New England epidemic of 1764 was so severe that the selectmen ordered guards around infected houses. This technique, which would later be adopted successfully elsewhere, did not work in Boston, and the epidemic spread, forcing the city to withdraw its guards. The death toll was not high. Almost five thousand people were inoculated, only forty-six dying of the operation. One third to one fourth of the population of the city voluntarily inoculated themselves.

Cities to the south of Boston were also smitten regularly after the 1721 episode. Philadelphia was hit in 1731; that epidemic soon spread to New York.

> Here is little or no Business, and less Money; the Markets begin to grow very thin; the Small-Pox raging very violently in Town, which in a great measure hinders the Country People from supplying this Place with Provisions. . . . The Distemper has been a long time very favourable, but now begins to be of the confluent kind, and very mortal. . . .

Between September 27 and October 11, 138 people died of smallpox in New York, almost 2 percent of the total population. The New York *Gazette* reported:

> In the Month of August last the Small Pox began to spread in this City, and for some Weeks was very favourable, and few died of this Distemper, but as soon as we observed the Burials to increase, which was from the 23rd of August, in our Gazette, No. 305, we began to insert weekly, the Number both of Whites and Blacks that were buried in this City, by which account we find, that from the 23rd of August to this Instant [mid-November] which is two Months and three weeks, there was buried in the several burying places of this City as follows, viz.
>
> Whites in all 478
> Blacks in all 71
> Total 549

Eventually 7 percent of the population would be buried and half sickened. "The living are scarcely able to bury the dead, whole families being down at once, and many die unknown to their neighbours."

A missionary in Burlington, New Jersey, northeast of Philadelphia, wrote:

> Soon after my Removal to Burlington that deplorable Distemper ye Small Pox raged among the people and Spread farr and Near, that for Six Months Upwards ye Congregation was very Small, partly by the mortality that followed that infection and the fears that ye generality of men were under of going abroad, Even to ye Publick worship of God.

The epidemic naturally ravaged the Indians, and the Senecas began dying in flocks. An epidemic in Canada forced many Indians there to cross the border, stirring up fears of more disease.

About half the population of Charleston, South Carolina, was felled by the 1738 outbreak despite the inoculations of Kirkpatrick. More were stricken in the surrounding countryside. Half the Cherokees were wiped out. They blamed the English and threatened to take their trade to the French.

The disease returned in 1745, again beginning in New York. Official condolences were sent to the Onondaga Indians, who were being decimated. The epidemic spread south into Delaware and Maryland, some of it "malignant." It came back in 1755. Its only beneficial effect was to prevent some Indian wars. A French dispatch revealed that although the Indians along the New England-Canadian border were ready to go on the warpath, smallpox sapped their strength.

Further condolences were sent to the Indians in Pennsylvania, and a meeting between Indians and whites scheduled for Philadelphia was postponed when one of the Indians broke out with the pox. The Indians were all sent home quickly.

The Maryland Assembly was postponed for five months when the disease struck Annapolis. Only fourteen members showed up, half of what was needed for a quorum.

To the south the Indians of Georgia and South Carolina were worse than decimated. Half the Catawbas died, many throwing themselves into rivers in agony. Others fled into the woods to die. The Indians promptly passed the disease back to the whites. During the signing of the Treaty of Fort St. George between South Carolina and the Cherokees, infected Indians spread the disease among the soldiers.

> As few of his little army had ever gone through that distemper, and as the surgeons were totally unprovided for such an accident [the Governor's] men were struck with terror, and in great haste returned to the settlements, cautiously avoiding all intercourse one with another and suffering much from hunger and fatigue by the way.

The soldiers, of course, brought the disease back to the city and Charleston reported another epidemic in 1760. Garrisons of soldiers who were to protect the settlers were cut down by the disease. Inoculation did not help. On March 20 the South Carolina correspondent for the *Pennsylvania Gazette* wrote:

> 'Tis to be presumed that you will naturally expect some news relative to the present Situation in this Colony, which you will, in a few words, when I assure you, that no Description can surpass its Calamity— What few escape the Indians, no sooner arrive in Town, than they are seized with the Small-Pox, which gen-

erally carries them off; and, from the Numbers already dead, you may judge the Fatality of the Disease. Of the white Inhabitants 95; Acadians 115; Negroes 500, were dead two Days ago by the Sexton's Account. About 1,500 White Inhabitants, 1,800 Negroes, and 300 Acadians, have had the Distemper, and chiefly by inoculation.

As soon as it seemed that the epidemic would abate new ships arrived in Charleston to rekindle the tragedy. According to a conservative estimate, 9 percent of the population of Charleston may have died.

The role of the army in spreading the disease was so obvious that people soon began fearing soldiers. Albany, New York, was declared off limits to troops camped outside.

In Philadelphia a society was formed to inoculate the poor, the practice generally being restricted to the middle class until that time.

Inoculation was, as we have seen, not an unmixed blessing. There were a number of tragedies arising from the operation. On August 14, 1767, an agent in Virginia, William Nelson, wrote to his principal in England:

> Our country man Mr. Smith arrived in high Spirits: hath set up his hospital for Inoculation at Flat's Bay; and proposes to begin his business as soon as the weather grows cooler; But some People object to his bringing the infection to a Country or Neighborhood that is free from it; If it comes by Chance, then let him begin and prosper as fast as he pleases. However, he goes to Baltimore soon with some of his family; wch. he will begin upon there and bring with him matter enough to infect the world, a second Pandora's Box.

[Smith didn't leave town fast enough. In February

> Nelson wrote]: Mr. John Smith hath rendered himself very blamable and unpopular suffering some of his Patients to go abroad too soon: so that the Distemper hath spread in two or three Parts of the Country: some of the College youths carried it to Wmsburgh where two or three have died, but by the care of the Magistrates it is stopped.

By 1790, in fact, all of New England had banned variolation, just in time for Jenner's vaccine.

During the Revolution the disease was again spread by the warfare, and effects of the disease on the outcome of the war are detectable.

During the meeting of the Continental Congress in Philadelphia in 1775, the physicians of the town agreed to stop all inoculation because some of the delegates, particularly those from the south, had never had the disease and there was some concern that they might be too susceptible to cases begun by inoculation. As the war progressed the disease became endemic and that concern irrelevant.

George Washington ordered all his troops inoculated and set up a special hospital for the purpose. But his efforts did not prevent the disease from once more saving Quebec City.

On August 20, 1775, Washington ordered General Philip Schuyler to plan an invasion of Quebec. It would match the attack on Montreal and force the British to fight a double-pronged assault. "I can well spare a detachment for this purpose of one thousand or twelve hundred men," he wrote. He suggested an overland route, which proved to be a disaster.

The detachment put together totaled a little over a thousand men and included such forces as Morgan's Rifles of Virginia and the nineteen-year-old Aaron Burr. It was placed under the command of General Benedict Arnold.

Arnold, using "battoos" (batteaux, or barges), sailed up the Kennebec River of Maine and then began a series of long, difficult portages. The folly of the overland route proved itself quickly. Soon the army was battered by the weather, by the rugged terrain and by a lack of supplies. It was quickly reduced to eating its shoes.

It was a famous march and it ended in disaster.

The American forces were so reduced by the journey that they did not have the strength to take Quebec but were compelled to lay siege to the city and dig in for the winter. Washington asked the colonies for more troops and three, Massachusetts, New Hampshire and Connecticut, agreed.

Among those who joined the expèdition from Connecticut was a young physician, Lewis Beebe. His journal, published in 1933 by the Pennsylvania Historical Society, is the best description of what happened.

Beebe arrived at the headquarters camp just after the first outbreak of smallpox. He was intrigued by what he saw. "Human nature," he wrote, "is acted out to perfection, in a variety of instances. The camp is one of the finest Schools in the world."

> On Thursday, General orders were given by Gen. Arnold for Inoculation, accordingly Col. Porters' Regt. was inoculated. On fryday, Gen. Thomas arrived at head Quarters from Quebeck and gave Counter orders, that it should be Death for any person to inoculate, and that every person inoculated should be sent immediately to [the American base at] Montreal. . . . General Thomas this day is under great indisposition of body.
>
> MONDAY, May 20th. General Thomas remained poorly, with many symptoms of the small pox. . . .
>
> TUESDAY, 21 May. Early this morning I was called upon to visit General Thomas, who upon first sight

> evidentally had the small pox, he thought it most adviseable to remove to Shamblee [Chambly] till the disorder should have its operation, he desired me to accompany and attend him. . . .
>
> SUNDAY 26 . . . If ever I had a compassionate feeling for my fellow creatures, who were objects in distress, I think it is this day, to see Large barns filled with men in every height of the small pox and not the least thing, to make them comfortable, was almost sufficient to excite the pity of Brutes.

Beebe left Chambly for St. John's where his regiment had moved. He had learned thoroughly to detest Arnold, whom he called an "infamous, villanous traitor" and wrote that he hopes future generations would see him as such. This was two years before Arnold actually became an "infamous, villanous" traitor.

General Thomas, who had seemed to be getting better on the 24th, suffered a sudden relapse. The man who had prevented inoculation of the troops died "a little after the first dawnings of the day."

> This was the 13th day after the eruptions first appeared.
>
> Thomas is dead that pious man,
> Where all our hopes were laid
> Had it been one, now in Command,
> My heart should not be grievd. . . .
>
> WEDNESDAY (June 5). For 10 days past I have been greatly troubled with the dysentary, and for three days it has been very severe, took Physic in the morning. Hope for some relief. In the afternoon went across the river to visit Col. Reed who I found to have the

disorder very light, the number of sick with the Pekot on this side is about 300, the greater part of which have it by inoculation and like to do well. . . .

FRIDAY 7. Last evening one died of the small pox, and early this morning one of the Colic, at 10 A.M. one of the Nervous fever, here in the hospital, is to be seen at the same time some dead, some Dying, others at the point of death, some Whistleing, some singing and many Cursing and swearing. . . . Visited many of the sick in the hospital—was moved with Compassionate feeling for poor Distressed Soldiers, when they are taken sick, are thrown into this dirty, stinking place, and left to take care of themselves. No attendence no provision made, but what must be Loathed and abhorred by all both well and sick.

On the 10th Arnold (the "great Mr. Brigadier General" to Beebe) ordered the regiment to move. "Is not this a politick plan, especially since there is not Ten men in the Regt but what has either now got the small pox; or taken the infection." The disease was raging through the army and on the 16th they had to load fifteen or sixteen batteaux with sick to send them from the front line back across the river to the hospital on an island. The next day Beebe returned to the hospital; his description of what he saw explains the failure of the Arnold expedition as well as anything can.

. . . [I] was struck with amazement upon my arrival, to see the vast crowds of poor distressed Creatures. Language cannot describe nor imagination paint, the scenes of misery and distress the Soldiery endure. Scarcely a tent upon this Isle but what contains one or more in distress and continually groaning, and calling for relief, but in vain! Requests of this Nature are as little regarded, as the singing of Crickets in a Summer's evening. The most shocking of all Spectacles was to see a large barn Crowded full of men with this disorder, many of which could not See, Speak, or walk–one nay two had large maggots an inch long, Crawl out of their ears, were on almost every part of their Body. No mortal will ever believe what these suffered unless they were eye witnesses. . . .

[More batteaux arrived with more sick.] WEDNESDAY 26. The Regt is in a most deplorable Situation, between 4 and 500 now in the height of small pox. Death is now become a daily visitant in the Camps. But as Little regarded as the singing of birds. It appears and really is so that one great lesson to be learnt from Death, is wholly forgot; (viz.) that therein we discover our own picture; we have here pointed out our own mortality, in the most lively colours. . . .

The remainder of the army eventually was withdrawn. The heat made it particularly unbearable to those with smallpox, Beebe reports, and there was no relief in sight.

SUNDAY 30. I hardly know what to say, I have visited many of the sick. We have a great variety of sore arms and abscesses forming on all parts of the body, preceeding from the small pox, occasioned by the want of physic to cleanse the patients from the disorder. However we had none so bad as yet, but what we have been able to cure except the disorder otherwise was too obstinate. Buried two today. No preaching or praying as usual. . . . A number are employed on the other side [of the river], almost the whole of the day to dig graves and bury the dead.

WEDNESDAY (July 3) . . . How strange it is that we

> have death sent into our Camp so repeatedly, every day? And we take so little notice of it? . . .
>
> FRIDAY 19. Last evening we had one of the most severe showers of rain, ever known; it continued almost the whole Night, with unremitted violence; many of their tents were ancle [*sic*] deep in water. Many of the sick Lay their whole lengths in the water, with one blankett only to cover them. One man having the small pox bad, and unable to help himself, and being in a tent alone, which was on ground descending; the Current of water, came thro his tent in such plenty, that it covered his head, by which means he drowned, this is the care that officers take of their sick. . . . Buried two yesterday and two more today. . . .

On July 22 Beebe set out with the sick for Fort George in the south. He found the hospital there contained two thousand sick. Presumably most had smallpox.

The disease killed one third of Arnold's army. Eventually he gave up and the expedition was withdrawn.

General Thomas died knowing the army was beaten. He described it as a "retreating army, disheartened by unavoidable misfortunes, destitute of almost everything necessary to render lives comfortable or even tolerable, sick and (as they think) wholly neglected and [with] no prospect of speedy relief."

Smallpox was a major factor in the retreat. It is not overstating the case to say that without smallpox Quebec could have been American.

It is not the only time in the War of Independence that the disease played a role. Lord Cornwallis, in the south, had made a strenuous effort to touch off a slave revolt, but smallpox undercut him. Jefferson, whose Monticello suffered from a smallpox epidemic in 1781, wrote that had Cornwallis' efforts been just to free the slaves "he would have done right, but it was to consign them to inevitable death from smallpox and putrid fever then raging in his camp. . . ."

"I suppose the state of Virginia lost under Lord Cornwallis' hands that year about 30,000 slaves, and that of these about 27,000 died of the smallpox and camp fever," he wrote.

Smallpox had been a part of life in colonial America. In part because of inoculation, the disease seemed to peak by 1750, while it was still growing in intensity back in England. There was very little correlation between English epidemics and outbreaks in the colonies. Most of the episodes in America came from the West Indies, not Europe.

Since it had not become an endemic disease to any extent in America until late in the century, it struck more adults than in England. It is possible that this was a factor in the American acceptance of variolation, which met much less resistance in the colonies than it did in England. It is also true that the American version of the disease seemed somewhat less deadly than the European, perhaps the result of a different strain of the virus, of sparser population or of a higher standard of living and hygiene.

But the effects of the disease were still wide-ranging. Smallpox frequently disrupted legislatures and schools, often paralyzed the economy, decreased the Indian population to such an extent that it helped clear land sought by the colonists and many times caused manpower shortages.

The number of Americans who died from smallpox is hard to determine, but there have been estimates. According to one study between 12 and 14 percent of the people stricken died, and almost everyone was stricken at one time or another. One report to the Royal Society stated that "it has been ascertained that one in 14 of all that were born died of smallpox, even after inoculation had been introduced; and of persons taken ill of smallpox in the natural way, one in five or six died."

Yet some good can also be traced to the disaster. A case can be made that the growth of American education owes its impetus to smallpox.

One French visitor to Virginia's William and Mary College in 1702 explained:

> There are about 40 students there now. Before it was customary for wealthy parents, because of the lack of preceptors or teachers, to send their sons to England to study there. But experiences showed that not many of them came back. Most of them died of smallpox, to which sickness the children of the West were subject.

One other interesting sidelight to the American experience is the attitude of the public toward the epidemics. They seemed to show a better understanding of contagions than that held by the scientific and medical community. It was the general public who demanded quarantine laws and public health measures, not the doctors. The medical community traditionally felt that disease was more complicated than mere contagion and did not place much emphasis on quarantine.

Nonetheless in America the ground was prepared for Jenner's vaccine. While Jenner had considerable difficulty selling its benefits in his country, it would be more readily accepted in England's former colonies.

VIII

The Vaccine Wars

No sooner had Jenner published his book than several roaring debates and feuds sprang up simultaneously. Jenner was like an overworked fireman, running from conflagration to conflagration. He spent the rest of his life both promoting and defending his discovery.

He quickly established himself as the bona fide discoverer of the vaccine and opponents who pushed the candidacy of the cattle dealer Benjamin Jesty soon retreated.

But that was only the beginning of the debate. Jenner was harmed as much by his friends as by his enemies.

Jenner returned to London after his book was published and brought with him a quill full of cowpox virus from the arm of Hannah Excell, the seven-year-old girl. Jenner gave it to a friend, Dr. Cline, at St. Thomas' Hospital. Cline had a patient with a bad hip; he thought that the infection from the cowpox might be a beneficial diversion, so he infected the boy, who went through all the predicted stages of cowpox. A colleague at the hospital, Lister, was convinced that Jenner's vaccine worked as advertised and published an account of his experience. This was picked up in the London press and that, more than the book, was how the world at large found out about the vaccine.

Scoffers popped up immediately. So did scoundrels. One erstwhile supporter, a Dr. Pearson, published a self-congratulatory pamphlet which gave the distinct impression that he was the discoverer of the vaccine, not Jenner. Another supporter, Woodville, demonstrated vaccination at his Paddington hospital, got his quills mixed up and started a smallpox epidemic.

Jenner published a new, enlarged edition of his book to warn everyone to be careful with the vaccine. Not all took heed. Pearson started his own smallpox epidemic, and there was at least one death.

The debate over the vaccine had philosophical and theological overtones, mostly deriving from Thomas Robert Malthus, the great diviner of the principle of population. Malthus insisted that overpopulation and its inherent poverty were probably among the causes of smallpox. In his *An Essay on the Principle of Population*—which actually began as a debate with his father in 1797, about the time Jenner was hot on the trail of the cowpox connection—Malthus describes smallpox as "the most prevalent and fatal epidemic in Europe" and the "most difficult to account for." For one thing, he says, it seems to occur all the time and everywhere, weather or season having nothing to do with it. Poverty and overcrowding, however, are another story:

> I do not mean, therefore, to insinuate that poverty and crowded houses ever absolutely produced it; but I may be allowed to remark, that those places where it returns are regular, and its ravages among children,

particularly among those of the lower class, are considerable, it necessarily follows that these circumstances, in a greater degree than usual, must always precede and accompany its appearance; that is, from the time of its last visit, the average number of children will be increasing, the people will, in consequence, be growing poorer, and the houses will be more crowded till another visit removes this superabundant population.

Smallpox, then, is at least partially the result of overpopulation and also one of the ways God keeps down the number of the lower classes, although not necessarily the most effective. "The small-pox is certainly one of the channels, and a very broad one, which nature has opened for the last thousand years to keep down the population to the level of the means of subsistence," he wrote, but there were far more effective ones, such as starvation. He was also sure that if Jenner's vaccine worked, some other disease would take over the role of smallpox in this divine plan, and the mortality rate would always continue basically unchanged.

But, as he pointed out in the 1826 version of the essay (which he was revising and rewriting through his life), that is no excuse for pushing vaccination or trying to end the wretched conditions of the poor that lead to diseases like smallpox:

> Instead of recommending cleanliness to the poor we should encourage contrary habits. In our towns we should make the streets narrower, crowd more people into the houses, and court the return of the plague. In the country, we should build our villages near stagnant pools, and particularly encourage settlements in all marshy and unwholesome situations. But above all, we should reprobate specific remedies for ravaging diseases, and those benevolent, but much mistaken men, who

have thought they were doing a service to mankind by projecting schemes for the total extirpation of particular disorders.

Malthus, who invented the science of political economics, had fervent supporters in England, including his fellow ministers. His effects lasted far beyond his death in 1834. For instance, Malthusians defeated the law that would have provided funds for the Act to Extend and Make Compulsory the Practice of Vaccination, which had been passed in Parliament in 1853. Without funds to support vaccination and actually produce vaccine, the law was meaningless. It wasn't until 1871 that Parliament voted £20,000 for vaccination. Only a fraction of that was spent in Ireland, which experienced devastating smallpox epidemics in 1871, '72 and '73.

Vaccination, you see, was against God's law.

To revert to a somewhat earlier time, Jenner was getting requests for his vaccine from all over the world. One of the most impassioned was from India, beset, as always, with deadly tidal waves of the disease. Jenner thought of sending boatloads of people to keep a chain of cowpox infection going, but the Austrians beat him to it, shipping vaccine via Constantinople.

Letters of congratulation also rolled in. One came from Thomas Jefferson * and another came from the Boston physician Benjamin Waterhouse. Waterhouse had heard of the vaccine through a friend, John Coakley Lettsom, who had sent Waterhouse Jenner's book, Pearson's pamphlet and an account of one of Woodville's more careful experiments. Waterhouse was impressed.

Waterhouse described Jenner's discovery in a Boston newspaper, the *Columbian Centinel.* Practically no one paid

* When Jenner died of a stroke in 1823 he was the most honored man of his age.

any attention, in part because there was no cowpox in America, and in part because variolation was, by now, widely accepted. The idea that someone ought willingly to accept an infection from a barnyard animal probably didn't make Jenner's ideas any more popular.

Waterhouse took the written material he had obtained from England and presented it at a meeting of the American Academy of Arts and Sciences, whose president was John Adams, then also President of the United States. Adams and the others were impressed, Waterhouse reported.

Now it was time to try the vaccine for himself. He wrote to Jenner and, after several tries, got some vaccine that was still alive.

Waterhouse decided to experiment first with his own children, using the thread impregnated with lymph that Jenner had sent:

> The first of my children that I inoculated, was a boy of five years old, named DANIEL OLIVER WATERHOUSE. I made a slight incision in the usual place for inoculation in the arm, inserted a small portion of the infected thread, and covered it with a sticking plaster.

Eight days later the little boy began to go through all the classic signs of cowpox vaccination. Waterhouse remarked that the sore on his son's arm looked just like the second drawing in Jenner's book:

> Satisfied with the appearances and symptoms in this boy, I inoculated another of three years of age [Benjamin, Jr.], with matter taken from his brother's arm, for he had no pustules on his body. He likewise went through the disease in a perfect and very satisfactory manner. This child pursued his amusements with as little interruption as his brother. Then I inoculated a

> servant boy of about 12 years of age, with some of the infected thread from England. The arm was pretty sore, and his *"symptoms"* pretty severe.

Waterhouse also vaccinated two other of his children and (unsuccessfully) two more servants. Then he decided to take Jenner's original experiment one further step, which Jenner himself probably should have done. He explained his purpose in a letter to Dr. William Aspinwall, the physician of the smallpox hospital in Boston; it was to try to infect his household with smallpox to see if the cowpox (which acquired the name kine pox in those days) really provided protection. He wanted Aspinwall to perform the actual attempt so that the public would trust the results. Aspinwall agreed immediately:

> Of the three which I offered, the Dr. chose to try the experiment on the boy of 12 years of age . . . whom he inoculated in my presence by two punctures, and with matter taken that moment from a patient who had it pretty full upon in. He at the same time inserted an infected thread, and then put him into the hospital, where was one patient with it the natural way. On the 4th day, the Dr. pronounced the *arm* to be infected. It became every hour sorer but in a day or two it dried off, and grew well, without producing the slightest trace of a disease; so that the boy was dismissed from the hospital and returned home the 12th day after the experiment.
>
> One fact, in such cases, is worth a thousand arguments.

Soon Waterhouse was in the vaccination business, becoming the principal propagandist for the procedure as well as the only doctor with live vaccine in America. Others tried to get vaccine, with mixed results, and many others went into the quackery business as soon as it became obvious that there was a market for vaccine. Waterhouse wrote:

> During this period, viz., the autumn of 1800, a singular traffic was carried on in the article of *kine-pock matter,* by persons not in the least connected with the medical profession; such as stage-drivers, pedlars, and in one instance the sexton of a church. I have known the shirt sleeve of a patient stiff with the purulent discharge from a foul ulcer made so by unskilful management, and full three weeks after vaccination, and in which there could have been none of the specific virus; I have known this cut up into small strips, and sold about the country as genuine kine-pock matter, coming directly from me. Several hundred people were inoculated with this caustic morbid poison, which produced great inflammation, sickness, fever, and in severe cases *eruptions.*

He said that he never could convince the victims of that episode that they had not been truly vaccinated and were still vulnerable.

That was only minor compared to the incident in Marblehead. Waterhouse went to Marblehead, a port sixteen miles from Boston, to vaccinate. One of the patients was a friend of a Marblehead doctor, Dr. S., the other the ten-year-old son of another physician. For vaccine, he used lymph from the arm of a sailor, obtained by Dr. S. The sailor had assured Dr. S. that the vesicle was cowpox. In reality it was smallpox.

Besides the two whom Waterhouse vaccinated, Dr. S. used the lymph on all his children. The result was a terrible smallpox outbreak in the town which killed a number of the children.

Such accidents were common as vaccination became

something of a fad in the United States and as a large number of people set themselves up in business despite the fact that real vaccine was still very rare.

Waterhouse railed against the charlatans and quacks and jumped to the defense of vaccination whenever a mishap occurred or a misstatement was made.

He still wanted to perform one more test to prove that the vaccine worked. It would go much further than Jenner had dared. He petitioned the Boston Board of Health for permission to test the vaccine on a large number of children who later would be exposed to smallpox. The Board of Health agreed, somewhat to Waterhouse's surprise. On August 16, 1802, he vaccinated nineteen boys from the poorhouse in his office while members of the board watched. All went through the typical vaccination symptoms.

The boys later were shipped to a specially built hospital on Noddle's Island, near Boston's Long Wharf.

> On the 9th of November, 12 of the above children, together with one other, George Bartlett by name, who had passed through the cow-pox two years before were inoculated for smallpox at the hospital . . . with matter taken from a smallpox patient in the most infectious stage of that disease. The arms of these lads became inflamed at the incision in proportion to the various irritability of their habits, but not to a degree greater than what any foreign, virulent matter would have produced. The small-pox matter excited no general indisposition whatever, through the whole progress of the experiments, though the children took no medicines but were indulged in their usual modes of living and exercise, and were all lodged promiscuously in one room.

But even that was not sufficient proof for Waterhouse. He used the modern technique of a control group. In order to prove that the smallpox virus he had tested on the thirteen boys was viable, he took more of it and inoculated two boys who had had neither cowpox nor smallpox. Both came down with smallpox. He counted five hundred vesicles on one of the unwitting guinea pigs. Still he was not through.

> When these pustules were at the highest state of infection, the 13 children before mentioned were inoculated a second time, with recent matter, taken from said pustules, which said matter was likewise inserted into the arms of the seven other children, who were absent from the first inoculation. They were all exposed, most of them for 20 days, to infection, by being in the same room with the two boys who had the small-pox, so that if susceptible of the disease they must inevitably have received it, if not by inoculation, in the natural manner.

Physicians from all over Boston studied the experiment and watched it being performed. The vaccinated boys remained safe from the smallpox. No modern scientist could have done a better job. In fact, few would even have gone as far as Waterhouse did on ethical grounds. The proof that the vaccine protected against smallpox was now simply irrefutable.

Waterhouse merely lamented that it took three years to get the experiment done but blamed the delay on the normal opposition expected to new ideas.

On June 8, 1801, Waterhouse had written to Jefferson, both as President of the United States and in his capacity as President of the American Philosophical Society in Philadelphia, complaining that he had received too many requests for vaccine, frequently from people whom he did not know. Many of the requests came from Virginians. He told Jefferson that he was afraid that if amateurs or incompetent physicians botched the vaccinations it would interfere with the proper spread of the practice in the United States. Waterhouse said that therefore he had refused to send the vaccine south and had decided to provide it to Jefferson, who was more likely to know suitable recipients. Since Jefferson wanted to vaccinate his slaves as well as his family, Waterhouse was careful to send drawings showing what the vesicles looked like on black skin. He also sent a manual describing the best way to vaccinate.

Along with the letter was a thread that had been dipped in lymph and placed between two glass plates to seal it. The plates were enclosed in cardboard and further sealed to protect the thread from the summer heat.

Jefferson passed the package on to a physician he respected and asked Waterhouse for more. Because temperatures at Monticello were in the high 90s, he recommended that the viruses be placed in a glass vial immersed in water. A Dr. Wardlaw vaccinated Jefferson's entire family and neighbors. Lymph from the family was then sent to Washington, the first smallpox vaccine in the new capital. Jefferson, in fact, had started his own strain of vaccine, which provided the substance to most of the south for a few years. He wrote to Waterhouse:

> In planting the disease here [in Monticello] I imagine it will be as salutary as anywhere in the Union. Our laws indeed have permitted the inoculation [variolation] of the smallpox, but under such conditions of consent of the neighborhood, as have admitted not much use of the permission. The disease therefore is almost a stranger here, and extremely dreaded.

Jefferson chose Dr. John Vaughan as his agent to carry the vaccine to Washington and explained to him:

> In the course of July and August, I inoculated about seventy or eighty of my own family; my sons-in-law, about as many in theirs, and including our neighbors who wished to avail themselves of the opportunity; our

> whole experiment extended to about 200 persons. One only case was attended with sore arms which required common dressings. . . . I have great confidence therefore that I preserved the matter genuine, and in that state brought it to Dr. Gantt of this place on my return, from whom I obtained the matter I now send you, taken yesterday from a patient of the eighth day. . . . In my neighborhood we had no opportunity of obtaining variolous matter, to try by that test the genuineness of our vaccine matter; nor can any be had here or Dr. Gantt would have tried it on some of those on whom the vaccination has been performed.
>
> We are very anxious to try this experiment for the satisfaction of those here. . . . I am, therefore, induced to ask the favor of you to send me in exchange, some fresh variolous matter, so carefully taken and done up, as that we may rely on it; you are sensible of the dangerous security which a trial with effete matter might induce. . . .

He sent vaccine also to John Redman Coxe in Philadelphia. Coxe reported back to Jefferson that it was successful and that he was going to publish the results. On April 30, 1802, Jefferson wrote back, saying it was a splendid idea to publicize the procedure. "I think it an important object in such a work to bring the practice of inoculation to the level of common capacities," Jefferson wrote. The main considerations were to make sure that the vaccine was always good and that those practicing vaccination were careful to select lymph only when it was at its peak:

> Is there such a point of time? I thought from my trials there was, but more extensive observations are necessary to ascertain that, and what the true point of time is, Mr. Vaughan had asked me to permit my letter to him to be published. My objection to it was that I

> am not a medical man, that it would be exposed to just criticism and that the observations of the medical gentlemen themselves would soon furnish what was better. If, however the letter can be useful as a matter of testimony, or can attract the notice or confidence of those whom my political course may have happened to make me known, and thereby engage their belief in a discovery of so much value to themselves and mankind in general, I shall not oppose it being put to that use.

He asked Coxe to make sure that any errors in his letter be corrected before it was published. Coxe's report was sent to the Secretary of State to preserve it. Jefferson soon wrote back to tell Coxe that his own vaccine strain had run out:

> The difficulty of keeping up a constant succession of inoculated subjects and the uncertainty of success from matter which is not fresh, will probably expose every part of the United States to the accident of losing the matter and render it a thing (of) common interest to all the Medical gentlemen, while they possess it, to distribute it far and near in order to multiply their chances of recovering it whenever the accident of loss may happen to themselves. Presuming you retain the matter at Philadelphia, I have to ask the favor of you, on behalf of the faculty here, to send some by post under cover to me. . . .

Jefferson had turned the White House into the southern distributorship of smallpox vaccine. Waterhouse referred all requests from the south to the President.

It was Jefferson who first gave vaccine to the Indians. In April, 1802, he was visited by a group of Indians under the leadership of Little Turtle. Jefferson gave the Indians plows and other agricultural implements and spinning wheels.

Then he told Little Turtle how the "Great Spirit" had

"made a donation to the enlightened white men; first to one in England, and from him to one in Boston, of the means to prevent them from ever having smallpox, which had occasioned great fatality among that race." According to the *European Magazine,* the Indian was so impressed that he allowed himself and the others to be vaccinated by the Reverend Mr. Gantt, the chaplain of Congress. Jefferson gave them a supply when they left, along with instructions written by Waterhouse on its use.

Later that year fifteen more Indians came to the White House and Gantt vaccinated them as well.

Waterhouse, in the meantime, was getting himself involved in a controversy that diminishes by a little the luster of his heroism.

He set himself up in the franchise business. His first customer was Lyman Spalding of Portsmouth, New Hampshire. At Spaulding's insistence Waterhouse granted him exclusive rights for the vaccine in Portsmouth in return for one quarter of the profits. He explained to Zacheus Bartlett of Plymouth:

> I would just observe to you that I have made a contract with several practitioners in different states and counties and most of them have agreed to give me one-quarter of the neat profits, others have given me a specific sum. In none have I *less* than a half of the profits.

He would not give any vaccine to other physicians in Boston and at least one of them tried to get his own from Jenner. Another advertised in a newspaper for cowpox victims. Waterhouse was worried that someone might "steal it before the month expires, and then not in six months it may be diffused over Boston or it may not."

The first to break the monopoly was Dr. Thomas Manning of Ipswich, Massachusetts, but the monopoly was not seriously threatened until Dr. James Jackson of Boston returned from England with a thread of virus and four lancets filled with vaccine. He discovered that Waterhouse had been vaccinating all over Boston and that his own lymph had failed. He went to Waterhouse and asked for vaccine. Waterhouse stalled him. Jackson had sent to London for more vaccine, but he decided not to wait and obtained a supply from Manning, who gave him all that he needed "and refused all compensation."

The monopoly was quickly breaking down. Several doctors protested Waterhouse's stranglehold on the substance but soon were able to obtain their own supply, usually from Manning. Some no doubt took lymph from people Waterhouse had vaccinated. Waterhouse held on for as long as he could, and many felt that his attacks on "spurious" vaccine used by others were to discourage anyone else from profiting. However, Waterhouse never charged Jefferson for vaccine. He seemed interested only in the New England "market," and it is doubtful that he made very much money from the vaccine.

Waterhouse was not the only one to trip into controversy over Jenner's discovery.

The most controversial vaccinator was James Smith, a student of Waterhouse's friend, Benjamin Rush, at the University of Pennsylvania. After Smith graduated from the university he went into practice in Baltimore in the mid-1790s. In 1797 he made his first excursion into public health when a yellow fever epidemic swept the city. Smith was one of the two physicians called by the Baltimore board of health to run the quarantine camp at the outskirts of the city. The city did its best to keep news of the epidemic quiet—it was bad for business—but Smith wrote of his experiences to Rush, and Rush told some journalists, who promptly revealed the news in print. The Baltimore papers claimed libel, and the board of health, denying that there was any epidemic, told Smith to shut up.

That touched off the first of many battles between Smith and the local power structure that eventually would lead to his downfall. Three years later, for instance, during another yellow fever outbreak, he set up a hospice in his own home and accused the board of malfeasance of duty. The board fired back at him, whereupon he blustered that he was "determined to support the *truth* and defend the cause of injured humanity, tho thousands of you should rise up against it." He was not popular.

Nevertheless, Smith was appointed attending physician at the Baltimore city and county almshouse, and on May 1, 1801, he performed his first vaccination, on a seven-year-old girl, Nancy Malcum. Then he vaccinated a number of the other wards and wrote to Rush that he had "apparent & promising success." He invited the other physicians in the city to share his vaccine but none did. He published an account of his work in the almshouse, however, and the Medical and Chirurgical Faculty of Maryland endorsed vaccination.

Smith decided that some organization had to be formed to make sure that vaccine was available to all and that it was pure and administered properly. He won the approval of the mayor and twenty-two of his colleagues to set up a center in his home to vaccinate the poor free and others for a fee. The first organization of its kind in America, it inspired immediate copies across the country.

Smith was so caught up in the vaccine that he named his first-born Edward Jenner Smith. His horizons quickly expanded. By 1809 he was petitioning the Maryland Assembly to establish a statewide agency with him as its head for a fee of $1,000 a year. He thought that he could let others actually run the agency while he returned to private practice. The Assembly had a different idea: a $30,000 lottery to finance a state vaccine agency. Smith would be signed up as head for six years and could take his salary from whatever was left over after expenses.

Unfortunately, the lottery made only $12,797 the first year, and an attempt to extend it into Pennsylvania failed. Smith went to Washington to ask to be made the agent of the national vaccine repository for the entire United States. He said that he would perform the job for nothing, hoping that Congress would not take him up on that part of the offer. But on February 27, 1813, Congress authorized President James Madison "to appoint an agent to preserve the genuine vaccine matter, and to furnish the same to any citizen of the United States." It gave the agent free mailing privileges and permitted him to collect fees for his vaccinations if he could. Madison named Smith.

The law, which was years ahead of anything Jenner's England had passed, for a few years made James Smith the main source for vaccine in the United States. Virginia's legislature followed the Congressional action by naming Smith its agent with a $600 salary. Smith had about twenty subagents, some of whom he dispatched with vaccine into the western territories. He also sent vaccine to other countries in the hemisphere if they requested it. He supplied the American military during the later years of the War of 1812 and helped beat back several serious smallpox epidemics with his vaccine and his assistants.

Smith, however, kept trying for a more formal arrangement, including a fee for his services to the government, but bills setting up such an official agency died in Congress, partially because of Constitutional objections.

In 1816 Maryland withdrew its support, and in 1818 Virginia named one of its own citizens as state agent. Smith decided to form a private foundation, the National Vaccine Institution, which would, he hoped, have a federal charter and its own building in Washington. The director would be appointed by the President of the United States. Smith figured that he needed to raise $40,000 as seed money. In less than two years he had more than half that amount and the foundation actually started business. But the Senate killed his request for a federal charter. Buoyed by the continued flow of private moneys, he decided nevertheless to push onward.

Smith soon ran into very bad luck. In the late summer of 1822 smallpox, imported from Liverpool, struck Baltimore. One of the victims was a girl Smith had vaccinated as a child. Smith, unnerved, told reporters it proved that vaccination was not completely protective. When one physician suggested that perhaps revaccinations were required, Smith, still shaken, replied: "I do not know who would not rather have the inoculated smallpox at once—than suffer such everlasting and uncertain repetitions of vaccination as you now propose." Smith, the champion of vaccination, was now proposing a return to variolation!

If that wasn't enough to destroy public confidence, there quickly came the Tarborough affair. On November 1, 1821, Smith sent some vaccine and a private letter to Dr. John F. Ward, the National Vaccine Institution's subagent in Tarborough, North Carolina. Late in December Ward wrote that the "vaccine" had caused smallpox and that several people had died. Smith tried to explain away the incident with all kinds of scientific circumlocutions but his enemies suggested that he had simply sent Ward the wrong stuff.

It turned out that *someone* had sent Ward smallpox scabs instead of vaccine. No one knows who was responsible. Ward said that his packet did not contain any letter from Smith and that the vaccine was labeled "Variol." Smith wrote to the Speaker of the House of Representatives to explain what had happened, and the House immediately turned the Tarborough affair into a political issue. The result was a repeal of Smith's mandate. The following year the funds for the National Vaccine Institution ran out.

Smith kept trying to persuade Congress to set up another national vaccine repository and kept picking at the Tarborough affair in an effort to vindicate himself. He also continued his roaring battles with his old nemesis, the Baltimore board of health, which once responded with the question: "What motives can prompt Dr. Smith to hang about small-pox patients like a preying vulture?"

Smith died in 1841, his dream of a national vaccination center doomed. Undoubtedly he had saved hundreds of thousands of lives with his National Vaccine Institution and his agency, but his failure of faith in the Baltimore epidemic and the Tarborough affair convinced a number of physicians to go back to variolation and discouraged many, at their peril, from trusting vaccination. Smallpox remained endemic in the United States through the century.

Meanwhile, use of the vaccine was spreading rapidly across Europe with varying degrees of success, depending on how enlightened the local governments were.

It did not take long for some to think up imaginative and humane methods of using the vaccine. Few could match the originality and drama of the Real Expedición Maritima de La Vacina, the boatload of orphaned boys who finally brought vaccine to the ravaged Indians in Central America.

The story of the Royal Expedition began on July 19, 1802, when the Ayuntamiento, the ruling council of Santa Fé de Bogotá, asked the Spanish king, Carlos IV, for help. A smallpox epidemic was racing through the city and funds were needed urgently. The viceroy of Bogotá had permitted the council to build a new hospital with a six-hundred-peso lottery but flatly refused to spend any public funds on controlling the epidemic. There were things that could be done. Quarantine was understood. Variolation had been used in the epidemics of 1779-80 and 1797-98. The Ayuntamiento was not satisfied with the viceroy's order.

In December, 1802, Carlos sent a letter to the Council of the Indies, asking them to take action. In February, 1803, the council ruled that the viceroy had not been out of line and suggested that the king drop the matter, which he agreed to do. But Carlos, who usually is depicted in history as an incompetent boob, had a fascination with medicine, and he was genuinely concerned about the toll smallpox was taking in his colonies. He already had had his three children vaccinated and the vaccine had saved the life of a princess, María Luisa, who had been vaccinated just as the disease was about to break out. He even had read Jenner. He wanted the council somehow to get vaccine to the colonies.

The council asked José Felipe de Flores, who had variolated Indians in Guatemala in 1780, to come up with a plan. There was no cowpox in the New World.

The trick was keeping vaccine alive long enough to make the transatlantic crossing. The standard method of preservation was to run a silk thread through the thick virus-laden matter from the sores, the lymph, and let the virus dry on the thread, which then could be shipped off. Other methods were to preserve the lymph within goose quills or store it in glass vials. Such methods would not work for the weeks needed to sail from Spain to Mexico or South America.

One method of keeping vaccine alive, Flores figured, was to keep infecting a chain of children. Five days after vaccination the vesicle appears on the arm. Generally it is fairly small with raised edges and a sunken center. But on the eighth day, it reaches maximum size and is round and distended with lymph. On the tenth day, the vesicle contains the most lymph. If the lymph is harvested then it can be used to vaccinate many children. Ten days later they would be ready to pass on a third generation of lymph. Flores reasoned that by outfitting a ship with enough children to pass along the vaccine for the duration of the voyage, together with a few cows with cowpox and vaccine sealed in glass, something was bound to get across alive. He even got the Pope to agree to link vaccination with the baptism ceremony to overcome religious opposition.

The king, however, was unwilling to pay for such an expedition. Finally, one of his physicians pointed out that smallpox was such a disruptive force on the economy of the colonies, that it was well within the king's best interests to pick up the tab. Carlos reluctantly concurred.

When the time came to find someone to head the expedition, a fifty-year-old physician, Francisco Xavier de Balmis, volunteered. Balmis was something of an expert on vaccination and had a background in military medicine. He seemed a likely choice.

Balmis' plan was to leave Spain from La Coruña and head for Veracruz, Mexico, via Tenerife, Puerto Rico, Havana and Mérida, with a few stops in between. He wanted to take boys from the poorhouse near the port. He did not think that cows or vaccine thread would be necessary, but the council insisted on at least the vaccine in glass vials. He took twenty-two children aged three to nine.

Five boys were vaccinated from one boy whom Balmis himself had earlier treated, and, after considerable trouble, he put the expedition onto the 160-ton corvette, the *María Pita.*

During the voyage he inoculated two boys at a time, figuring that as children they would scratch themselves or otherwise spread the vaccine in their arms, and with two boys working at the same time, the odds were better that one of them would keep the vaccine active. Also he made sure that none of the other boys touched the two who had been vaccinated so that they wouldn't catch cowpox out of turn. Every eight or ten days he would vaccinate another two.

On February 9, 1804, he landed in San Juan, Puerto Rico. He was greeted royally but the visit quickly degenerated into a brawl. It turned out that vaccine had beaten him there by way of the Danish island of St. Thomas. Balmis, who had a high regard for his own value, was irate. He challenged Dr. Francisco Oller, the physician who had brought the Danish vaccine, to prove that it was the real stuff by trying to infect one of his patients with smallpox. Oller agreed, choosing his own son and the bishop-elect as guinea pigs. When the smallpox failed to take, Balmis called off the challenge and after four weeks crept out of San Juan with his tail between his legs.

By the time he reached Puerto Cabello, Venezuela, Balmis had only one boy left with live vaccine. He took twenty-eight more kids from the local orphanage and vaccinated them to reproduce the virus.

The expedition split at this point, with his assistant, José Salvany, heading south to Bogotá, Peru and Buenos Aires. Balmis headed north toward Mexico.

The Venezuelan administration was pleased with Balmis' efforts and helped meet his expenses. In three days he vaccinated 2,064 people and spread the vaccine across the country. He distributed pamphlets that he had written giving directions for continuing the chain. By April 29 he had vaccinated better than twelve thousand people and headed toward Cuba.

After a storm-tossed passage of the Caribbean, he arrived to find that Cuba already had the vaccine—the Puerto Rican strain. This time Balmis behaved, setting up a vaccination center of his own and helping establish a repository for the vaccine.

On June 17 he learned that Salvany's ship had been wrecked but no one was hurt and the vaccine was still good. The next day Balmis left for Mexico, where he discovered again that he had been preceded by vaccinators using material from Cuba. After some two to three thousand people had been vaccinated, the strain had run out. Balmis lost his temper once more. His main problem now, however, was that no one seemed to care. The only way he kept the vaccine alive was to force twelve soldiers to be vaccinated.

Then the governor compelled some Indian women to bring in their children:

> I would have lost the lymph from all those beautiful vesicles resulting from my 12 vaccinations on the 10th if the zealous Alcalde had not dragged in 20 women whose children were vaccinated only after a thousand entreaties. They then exclaimed loudly that they did not owe anyone anything. Others cried that if they did owe something they could not pay. All the while they asked why they had been brought to this terrible place. Then every single one immediately went to the closest apothecary to get an antidote for the poison that had been introduced into their children's arms. . . .

Still no one showed up for vaccination and the viceroy refused to order mandatory participation. Balmis established a repository and then headed for the interior. In Puebla he vaccinated 10,209 people and set up a vaccine board when a priest promised to produce fifteen boys every nine days to keep the virus going. The vaccine slowly spread.

After a year's delay Balmis went on to the Philippines and brought the first vaccine to Spain's Pacific colonies.

Balmis visited the Americas one other time to see if the vaccine was still passing well, but political turmoil prevented him from conducting many tests. He died in Madrid on February 12, 1819. In fact the vaccine was distributed throughout the colonies, saving hundreds of thousands of lives, the first blunting of the disaster for the poor Indians.

But the Indians still remained the most vulnerable population in the New World and, despite the best efforts of many, they continued to be decimated by smallpox.

Everyone tried to save them, including, as we have seen, Jefferson. On June 20, 1803, he sent a letter to Captain Meriwether Lewis, then with the First Regiment of Infantry among the Indians, advising that one of the ways he thought Lewis could earn their friendship was with vaccine: "Carry with you some matter of the kinepox. Inform those of them with whom you may be of its efficacy as a preservative from the smallpox; and instruct and encourage them in the use of it. This may be especially done wherever you winter."

The first batch Jefferson sent, however, didn't work. Lewis wrote back: "I would thank you to forward some vaccine matter, as I have reason to believe from several experiments made with what I have, that it has lost its virtue." It is not known if Jefferson complied.

But the government's efforts to vaccinate and revaccinate the Indians eventually worked in many areas, to the extent that after a while in some places the Indians were safer from smallpox epidemics than many of the whites who lived around them. The expense for the vaccination was borne by the government under an act passed by Congress on May 5, 1832, during the administration of Andrew Jackson. The Secretary of War was ordered to supply pure vaccine to the Indian agents for the program.

In the ensuing years surgeons were sent into Indian lands to perform the procedure but often were met with resistance by the Indians, except when epidemics broke out among the tribes. Then many of the recalcitrant gave in and permitted the vaccination.

In 1849 two physicians in Santa Fe wrote to the Indian agent there:

> It has occurred to us that humanity would be greatly benefited by a thorough vaccination of the Pueblos under your government. The fatal and loathsome scourge, for which vaccination is a specific, almost always in its periodical visits to this country makes its appearance in their villages, where as anyone will in-

form you, the mortality is frightful. Their confined and ill-ventilated apartments nourish and propagate the poison to such an extent that it is unsafe for the unprotected citizen to go within their atmosphere. We have been informed on creditable authority that small-pox or varioloid has not ceased to exist in one or the other pueblos for the last 20 years. . . .

Another expert, E. T. Denig, wrote in 1854:

> It is hardly conceivable how the smallpox among the Indians can be cured by any physician. All remedies fail. The disease kills a greater part of them before any eruption appears. We have personally tried experiments on nearly two hundred cases according to Thomas' "Domestic Medicine," varying the treatment in every possible form, but have always failed, or in the few instance of success the disease has assumed such a mild form that medicines were unnecessary. It generally takes the confluent turn of the most malignant kind which in 95 out of 100 is fatal. It appears to be the natural curse of the red man. . . . We have tried from year to year to introduce general vaccination with kinepox among them and have even paid them to vaccinate their own children, but they will not have it done to any extent, and the few who will, do it more to please us than to benefit themselves.

Even as late as 1882 the government was spending only $1,400 a year for vaccinating the Indians, a situation that did not change materially until 1900.

Tribes along the Columbia and Missouri rivers suffered most in the early nineteenth century. The plight of the Omahas is particularly tragic. The Omahas lived near Omaha Creek, a tributary of the Missouri. In 1802 they were hit by a double epidemic of cholera and smallpox. As usual, the worst suffering and disfigurement was among the children. So tragic was the effect on their children that the Omahas decided to commit glorious suicide in combat like true warriors. All the members, men, women and children, formed a huge war party and attacked some of the tribes that had been harassing them for years, the Ponca, the Cheyenne, the Pawnee and the Oto. The Omahas suffered huge casualties. Eventually they ran out of enemies and limped back to Omaha Creek.

In 1804 they were visited by Lewis and Clark. Clark wrote:

> Those people, having no houses, no corn, nor anything more than the graves of their ancestors to attach them to the old village, continue in pursuit of the buffalo longer than others who have greater attachment to their native village. The ravages of smallpox which swept off four hundred men and women in the population, had reduced this nation to not exceeding three hundred men and left them to the insults of their weaker neighbors, who had previously been glad to be on friendly terms with them. I am told when this malady was among them they carried their frenzy to every extraordinary length, not only burning their village but they put their wives and children to death with a view of their all going together to some better country. They bury their dead on the top of high hills and raise mounds on top of them. The cause or way those people took the smallpox is uncertain, the most probably, from some other nation by means of a war party.

The Omahas lost two thirds of their people, including their chief, Washinga Sakba ("Blackbird"). Blackbird, in accordance with his wishes, was buried in full war dress and bonnet on a bluff overlooking the Missouri on his favorite war horse—the horse was alive.

In several instances white men forced Indians off the land or forced peace by threatening them with smallpox. One trader imposed his terms by holding up a bottle and saying, "In this bottle I have it confined. All I have to do is to pull the cork and send it forth among you and you are dead men."

But perhaps no tribe suffered the way the Mandan did in 1837. It is a tragedy of good intentions and barbarism, and in the end an entire tribe was lost.

The story began in June, when the steamboat *St. Peter,* owned by the American Fur Company, left St. Louis to trade up the Missouri. Sixty miles above Fort Clark (near the modern city of Bismarck), a mulatto ship hand broke out in smallpox pustules. Others, including a Mr. Halsey, came down with the disease. It was the 24th.

According to the legend a Mandan chief stole a blanket from the steamboat. A trader, Francis A. Chardon, saw the danger and offered to pardon the thief and present new blankets for the one stolen. Whether the blanket was ever returned is not known and it is also unknown if it was really the means of transmission. Indians were in and out of the river fort constantly.

Meanwhile, those on board the ship tried to stem an epidemic. They took scabs from Halsey and tried to inoculate thirty squaws and a few white men, but most caught the disease and died. The victims were locked up inside the fort, half eaten by maggots.

The disease spread throughout the fort, so the next task was to try to keep it from spreading beyond the fort to the Indians. One day a group of forty Indians came to beg food. When they grew persistent someone in the fort opened the back door a crack and showed the Indians a boy whose face was swollen with smallpox. The Indians fled, but even that brief encounter was enough to kill half of them.

Runners were sent out to the Assiniboin to tell them to keep away, but that didn't work either and the disease spread with almost 90 percent mortality. Other tribes soon caught it. One trader came to a Piegan village and found only dead bodies. Some six thousand Piegans, Blackfeet and Bloods died.

Meanwhile, the Mandans continued to suffer. They were surrounded by enemies and unable to flee. They died trapped in their wigwams, the stench of the dead unbearable for miles.

An anonymous visitor to the Mandans wrote in June of 1838:

> The destroying angel has visited the unfortunate sons of the wilderness with terrors never before known, and has converted the extensive hunting grounds as well as the peaceful settlements of those tribes into desolate and boundless cemeteries. The number of the victims within a few months is estimated at 30,000 and the pestilence is still spreading. The warlike spirit which but a few months ago gave reason to apprehend the breaking out of a sanguinary war, is broken. The mighty warriors are now the prey of the greedy wolves of the prairie. . . .
>
> Among the remotest tribes of the Assiniboin, from fifty to one hundred died daily. The patient, when first siezed, complains of dreadful pains in the head and back, and in a few hours he is dead: the body immediately turns black, and swells to thrice its natural size. . . . For many weeks together our workmen did nothing but collect dead bodies and bury them in large pits; but since the ground is frozen we are obliged to throw them into the river. The ravages of the disorder

> were the most frightful among the Mandans, where it first broke out. The once powerful tribe, which by accumulated disasters, had already been reduced to 1,500 souls, was exterminated with the exception of 30 persons.

Some Indians were on hunting parties when the epidemic broke out and returned to their villages to find their children and wives dead. Many could not take the tragedy and committed suicide with their knives or guns. Some threw themselves off tall cliffs. The prairie for miles was a "vast field of death, covered with unburied corpses and spreading for miles, pestilence and infections."

The survivors roamed the prairie begging for food as the epidemic spread. The Assiniboin and the Blackfeet were nearly exterminated.

> No language can picture the scene of desolation which the country presents. In whatever direction we go, we see nothing but melancholy wrecks of human life. The tents are still standing on every hill, but no rising smoke announces the presence of human beings and no sounds but the croaking of the raven and the howling of the wolf interrupt the fearful silence.

As the disease spread it is believed some sixty thousand Indians died. It was almost forty years after the discovery of vaccination.

In 1845 the Crows conquered what was left of the Blackfeet, but just as the victory celebrations were to begin, smallpox hit the victors. The Blackfeet, survivors of the 1837 epidemic, were asked by the Crows how to withstand the disease. According to witnesses the Blackfeet cagily suggested ice-cold baths. The Crows did as they were told and the result was devastating to the victors, who died of shock.

From New York to Washington State the Indians continued to suffer from smallpox in the face of feeble vaccination efforts, which did not really expand until after the Civil War. There was almost no year in which some tribe was not being decimated. It was only with the dawn of the twentieth century that the disease finally abated and variola minor replaced variola major on the reservations, probably imported from South Africa.

While the Indians still were gravely menaced by smallpox, in urban regions of America and the more established areas of the east and midwest, vaccine quickly dulled the terror of smallpox and the disease became something less than a major concern.

There were only fourteen deaths due to smallpox in New England in the nineteen years before 1832. One Boston physician wrote:

> Since the introduction of vaccination, the student of medicine has been tempted to pass over the leaves of his text-book containing the details of this loathsome and fatal disease, as he does those relating to the plague, the black death of the 16th century or the sweating sickness; considering it as a form of epidemic, if not absolutely extinguished, as but little likely to come within the immediate sphere of his duties. . . .

That does not mean that epidemics ended, even in the largest cities. New York City was stricken with an epidemic in 1858 and 425 people died. In 1850 smallpox was the eleventh greatest killer in the United States, causing 2,352 deaths. (Cholera was the worst, taking 31,506 lives in one year.)

One of the problems was the difficulty of producing active vaccine. Getting live viruses from cows in the United States was very rare, so vaccinators had to depend on arm-to-arm inoculations. Soon, however, the Jennerian stock, as it was called—the viruses which stemmed from those first shipped over by Jenner—weakened and became useless. Some of it had gone through seventy generations, getting slightly weaker each time.

In April, 1866, there was a cowpox epidemic in Beaugency, France. The French government learned to keep the virus going through infection of heifers, and four years later this new stock of vaccine, the Beaugency stock, was brought to America and replaced the Jennerian.

The system was bound to be improved. In 1834 Negri in Naples produced good quality calf lymph by infecting healthy cows, making arm-to-arm inoculations unnecessary, although many physicians preferred to stay with the old way. The method was introduced into France in 1864 and to Prussia and Brussels two years later. H. A. Martin of Boston was the first to distribute calf lymph in the United States in 1870. He stored it in a quill. The dried vaccine, however, did not live very long and it was years before calf lymph, improved by the addition of glycine, replaced arm-to-arm vaccinations.

Essentially Negri's method is still used today. Vaccine is produced by shaving the belly and inner thighs of a calf and scratching vaccinia viruses into the skin. In about eight days the crusts of the pox marks can be harvested. The vaccine then is freeze-dried, which scientists believe keeps it viable forever. All that has to be done is to add a wetting solution, usually glycine, to reconstitute the viruses in the vaccine.

Even in the 1880s vaccination had to be defended against the stubborn. A Philadelphia physician, Joseph F. Edwards, published a monograph in 1882 defending vaccination. He said that criticism was based on two grounds. One was that there was always a risk that other diseases could be spread by vaccination. Indeed there were cases in which careless vaccinators infected their patients with other diseases, particularly syphilis. The other ground for attack was the irrefutable fact that vaccination did not always work.

The problem was not with the principle of the vaccine so much as with the vaccinator. Not every physician was competent or honest, he admitted. One physician was asked to perform a vaccination, Edwards wrote, but he had no vaccine. "Fearing to lose his fee," he inoculated the patient with gum arabic. He told the patient to come back in a week if the vaccination didn't take.

Of course it did not take and the patient had to return. By that time the physician had the real thing and obtained his second fee.

Edwards also warned that there was a Philadelphia doctor who was buying scabs from the Municipal (Smallpox) Hospital, mixing them with water and selling them out of town as "Genuine Bovine Virus Direct from the Cow."

In 1881 the British humor magazine *Punch* parodied:

> To vaccinate or not, that is the question!
> Whether 'tis better for a man to suffer
> The painful pangs and lasting scars of smallpox,
> Or to bare arms before the surgeon's lancet,
> And by being vaccinated, end them. Yes!
> To see the tiny point, and say we end
> The chance of many a thousand awful scars
> That flesh is heir to—'tis a consummation
> Devoutly to be wished. Ah! soft, you, now,
> The vaccinator! Sir, upon thy rounds
> Be my poor arm remembered.

Vaccination in the West became a standard public health procedure although many people refused to submit themselves. As late as 1892 there were 13,899 cases of smallpox in Jenner's England, with 1,182 deaths. London during that period recorded 4,759 cases and 381 deaths.

But all rational scientists knew that vaccination worked. They did endless studies to prove it. A Dr. Schuyler at the hospital in Troy, New York, reported in 1882 that of 199 cases, seventeen people had been vaccinated at least once in their lives and two of them had died. The other 105 had never been vaccinated and thirty-three died. No one who had been vaccinated recently had the disease.

Another study in Sweden showed that in the period between 1774 and 1801, when vaccination was unknown, there were 1,973 deaths per million in Sweden. In 1802-1816, when vaccination was known but not obligatory, there were 479 deaths per million. Between 1817 and 1879, when it was mandatory to be vaccinated, Sweden's rate slipped to 181 deaths per million. Not every one was vaccinated, apparently.

There were still tragedies. During the Franco-Prussian War the French, who did not believe in revaccination, suffered terribly from smallpox, sending their sick soldiers back to Paris in cattle cars. The Germans, who revaccinated routinely, barely noticed the disease.

But the disease was, by no means, done with yet. There was the rest of the world to deal with.

IX

The Twentieth Century

Great Britain celebrated the turn of the modern century with an old-fashioned smallpox epidemic. A hundred years after Edward Jenner's remarkable discovery, his native land was still vulnerable. The British had never become used to the fact that one vaccination did not provide lifelong immunity, and most Britons simply were unprotected.

What made the epidemic of 1900-02 odd was the fact that it arrived in England by mail. News accounts said the virus was traced to a parcel shipped from Salt Lake City, Utah, to a Mormon missionary in London.* By the time the epidemic was spent, there were nearly a thousand deaths. According to the official records there were 2,278 cases in London

* Researchers have found that there was a Mormon convention in London that year and the disease was probably carried by a conventioneer. It happened again in 1913.

among unvaccinated people and 753 of them died, a mortality rate of 33 percent.

The rest of Europe was not in much better condition. In the first two or three years of the century, Paris reported 4,505 cases and 758 deaths, Italy had 60,532 cases and 14,951 deaths, and more than 1,500 died in Spain. Variola minor was found in most of the continent and variola major around the Mediterranean basin.

The problem, besides the failure to enforce vaccination, was transportation. Too many people and too many goods were being transported across oceans for the disease to be isolated. In 1901 Meridith Young, a physician in Stockport, England, claimed to have traced a very mild case of smallpox to a bale of cotton shipped from Texas.

Even in Jenner's beloved countryside, smallpox was still something of a constant threat. Between 1902 and 1905 there were 1,305 cases of the disease and 247 deaths in the English provinces. Most were undoubtedly among the unvaccinated.

While most of the smallpox in the United States was variola minor, the killer type was hardly unknown. New York City suffered an epidemic of it in 1901-02, with 3,480 cases and 720 deaths.

There were 48,206 reported cases in the United States in 1901, with 1,085 deaths (a mortality rate of 2.4%); in 1902 there were 55,857 cases and 2,111 deaths (3.7%); and in 1903 there were 40,581 cases and 1,382 deaths (3.4%). The disease was always present some place in the country, but the vaccine kept epidemics to a minimum. Shortly after the turn of the century variola major virtually disappeared and most of the reported cases were variola minor.

A breakdown of the states also shows that most of the cases were in rural or frontier areas. There were 25,598 cases reported in 1901 (note that not all states made a report):

STATE	CASES	STATE	CASES
California	177	New Jersey	22
Colorado	1,096	New York	353
Connecticut	0	North Carolina	4,281
District of Columbia	96	North Dakota	306
Florida	1,286	Ohio	750
Illinois	730	Oklahoma	2,342
Indiana	764	Oregon	164
Iowa	850	Pennsylvania	168
Kansas	2,202	South Dakota	365
Maine	8	Texas	2,925
Maryland (exc. Baltimore)	6	Utah	966
Massachusetts	144	Vermont	0
Michigan	2,585	Virginia	350
Minnesota	1,002	Washington	583
Montana	634	Wisconsin	443

Variola major was reported in Texas, Kansas and California in 1911. The strain struck the last two states two years later and was also reported in Oklahoma and parts of Pennsylvania. The city of Pittsburgh had 121 cases and 33 deaths.

But it was obvious that the widespread use of vaccine and the enforcement of public health laws had already turned smallpox into a minor health threat in the United States. In 1900 there were 0.3 deaths per 100,000 people caused by the disease. In 1910 it was 0.5 per 100,000, and by 1920 it was down to 0.1.

The same was true in most of Europe as the century progressed. There were exceptions, the most important being Russia. There were fifty to a hundred thousand cases reported there every year, and no one knew the real extent of the disease.

World War I led to an inevitable upsurge in smallpox as

well as in every other communicable disease. The greatest vector was the unvaccinated Russian prisoners of war in Germany. In 1917 they touched off an epidemic that in Berlin alone occasioned four thousand cases and four hundred deaths. The disease even spread to peaceful Sweden. In Russia, racked by the revolution, there were 186,000 reported cases in 1919 and, probably, an unthinkable number unreported. Some countries bordering or near Russia also were affected: Rumania, 20,523 cases, Yugoslavia, 5,278, Czechoslovakia, 11,209, and Germany, 5,012.

The Mediterranean area continued to suffer. More than 3,500 died in Spain.

But when the western world finally settled down by 1920, smallpox began to diminish markedly. The Bolshevik government in Russia instituted mandatory vaccination by 1923, ending the dark days of the disease.

By 1928 smallpox had disappeared from many countries in Europe, and was rare in the others. On the eve of World War II there were only 750 cases of smallpox in all of Europe, all but 50 of them in the Iberian peninsula and the Soviet Union. There were fewer than 600 cases a year in the United States.

Sometime around 1920 variola major virtually had disappeared even in those places in Europe where the disease still existed. From then on almost every case, except those due to importations from the still endemic parts of Asia, Africa and South America, was variola minor. One disadvantage was that when the disease struck, it was taken less seriously. For instance, England had a smallpox epidemic that lasted twelve years, from 1922 to 1934. Because the infection was mild, the precaution of mass vaccination was not adopted as in all earlier outbreaks. So in that period there were 81,249 cases, 14,767 in 1927 alone. This 120 years after Jenner!

But by no means was the disease gone from the consciousness of the Western world. There were constant importations from the festering remainder of the world, and no place was safe.

In March, 1947, an American businessman, Eugene LaBar, who had lived twenty-two years in Mexico City, stepped off a bus in New York City. He and his wife were on their way to retirement in Maine, but LaBar had to get off. He wasn't feeling very well.

The couple checked into a Manhattan hotel until he would feel better. A week later, however, his fever was worse, and he went to Bellevue Hospital. Doctors suspected bronchitis and hemorrhages, and he was transferred to Willard Parker Hospital for Contagious Diseases. Two days later he died. The diagnosis was still bronchitis.

There were several patients LaBar came in contact with while at Willard Parker, including a twenty-two-month-old-girl with tonsillitis and a twenty-five-year-old Puerto Rican, Ismael Acosta, who had mumps. The baby and Acosta were subsequently released.

But on the 21st of March, both returned to the hospital with symptoms that looked suspiciously like smallpox. By tracing their records the doctors established the contact with LaBar. They sent samples taken from all three to the U.S. Army Medical School Laboratory in Washington, and on Friday, April 4, the doctors were informed that all three samples showed smallpox. It was the first time since 1939 that any case had been reported in the city, and that was an importation with no secondary infections. There had been two cases in 1922.

Despite the danger of the disease and the rarity of the event, the smallpox report got very little initial attention. The *New York Times* placed it on page 21 in its Saturday, April 5, edition, along with a statement by Health Commissioner Israel Weinstein that the chances of an epidemic were "slight." Weinstein nevertheless urged revaccination for everyone in New York City who had not been vaccinated since childhood. The city health department went searching for contacts of the three victims (there were "several thousand," Weinstein reported) and when found vaccinated them. Vaccination was to be free at all health district stations and at outpatient clinics in the municipal hospitals. The two patients, Weinstein said, were "doing well."

New Yorkers were distracted by enough other things not to worry over smallpox, a disease they knew little about and probably had ceased to fear from lack of experience. In Centralia, Illinois, that week there had been a terrible coal-mine disaster that killed 111 miners, and the telephone workers were set to stage their first nationwide strike. President Harry S. Truman was considering sending the Army in to take over the Bell system and keep communications going. And, obviously, World War II had been over less than two years, and a lot of people were trying to get their lives in order. It is small wonder that Eugene LaBar's bad luck did not interest most of them.

Yet a few people read about the three smallpox cases and lined up for vaccination the next day, Sunday. Health workers were put on three shifts for twenty-four-hour processing of vaccine in the health department's lab. Health centers stayed open until 8 p.m. Hospitals were asked to have vaccine always available in case someone popped in at any hour to request vaccination.

All twenty thousand employees of the municipal hospitals were vaccinated. Weinstein said that he was gratified by the public's response. Deputy Health Commissioner Samuel Frant assured one and all that "there is sufficient vaccine available for everyone."

On Monday, April 7, there was another scare: a young

Cuban who was transferred to Willard Parker with either smallpox or chickenpox proved to have the latter.

Yet even when Acosta's wife, Carmen, became the fourth victim of the disease, the city did not take much notice. About fourteen hundred people in her Harlem neighborhood were vaccinated, along with another several thousand city workers. The city asked druggists to purchase vaccine from commercial sources so that it would have enough for its free program. The search for contacts spread across the country as epidemiologists tracked down people on LaBar's bus and everyone who stayed in his midtown hotel.

On the weekend of April 12-13, however, the populace finally realized that it had the beginnings of an epidemic on its hands. Carmen Acosta died. A nun and three children at the Cardinal Hayes Convalescent Hospital for Children in Millbrook, New York, seventy-five miles north of the city, came down with smallpox. The disease had been brought to Millbrook by a child who had been at Willard Parker. Everyone at the Catholic institution was vaccinated and everyone in the little town of Millbrook also demanded vaccine, which immediately ran out.

Mayor William O'Dwyer of New York City was told about the problem and was urged by Weinstein to take drastic action. The health department wanted to vaccinate everyone in the city, all 7.4 million of them, a seemingly impossible task but one that Weinstein thought necessary if the epidemic was to be stopped quickly. O'Dwyer agreed. He had himself vaccinated by Weinstein and then held a news conference to announce that the situation, "while not unduly alarming, was quite disturbing." He announced a plan to vaccinate everyone within three weeks and turned over city facilities to the free vaccination program, including all 84 police stations and 383 firehouses.

"The time has come to urge everyone in the city to get

vaccinated against smallpox," O'Dwyer told reporters. "It is a virulent disease; with a 40 percent mortality among children and 20 percent among adults. It is a very big job but it should be done. Many vaccinations made years ago are no longer effective. Many persons have never been vaccinated. Air travel has increased the danger from this disease.

"I have consulted with Dr. Weinstein and Dr. Thomas M. Rivers of the Board of Health, a leading world authority in this field. They think all the city's available medical machinery and manpower should be thrown into this situation to vaccinate everyone in the city within the next three weeks. There is no need for panic or undue alarm, but there is need for every precaution as soon as possible."

All 7.4 million New Yorkers would not, in fact, be vaccinated. But even those who proposed the plan were probably amazed at how close they came. The terror of smallpox was not hard to revive. The *Times* was impressed enough to put the story on page 1 for the first time, and nothing makes an emergency official better than that.

It was estimated that by Sunday, April 13, forty-two thousand, mostly municipal workers, had been vaccinated. The city had a long way to go.

The next day, when the fifth case was reported, the city realized that it did not have enough vaccine. O'Dwyer asked Secretary of War Robert Patterson and Secretary of the Navy James Forrestal for help. The request was granted. A few minutes after the telephone call to Forrestal 250,000 doses were transferred from the Brooklyn naval supply depot to the health department labs on the East River Drive. The Army checked its stores and also promised supplies. Both services volunteered medics for the vaccination program, but the city, possibly worried about lawsuits should something go wrong, politely declined. Dr. Ralph S. Muckenfuss of the health department said he thought that five million doses could be found within ten days.

Mobilization of city workers continued. In an emotional speech Police Commissioner Arthur W. Wallender told his officers: "It is a task for you to get serious about. It may become one of the worst emergencies that we may be confronted with and needs our attention at once."

Most of the air raid wardens volunteered for duty. The public responded. Police had to control lines at the Arthur Avenue health center in the Bronx.

The state police were called into Millbrook to help contain the growing panic there. Two thousand villagers and farmers were vaccinated.

Every radio station in the city blared out warnings: "Be sure, be safe, get vaccinated!"

The city dispensed 634,421 doses on the 15th as the number of known local cases jumped to eight. Weinstein said that he thought all the contacts and possible cases now had been traced and doubted that there would be any others, but he urged that the vaccination program continue. Half a million doses were flown in by the Army from Moffett Field, California, and the city purchased 2 million doses from commercial sources.

New Jersey, across the Hudson River, began its own smallpox campaign despite the lack of evidence that cases had crossed the river. Sweden ordered all travelers from the United States vaccinated and the health commissioner of Westchester County complained that New York City had "cornered the market on vaccine."

On the 16th the vaccine ran out. People had lined up and waited for hours, sometimes in the rain, for vaccine and the city, which had been vaccinating at the rate of 200,000 a day, had finally exhausted its supply. The problem was the private suppliers. So far 1,222,540 doses had been distributed, but only 42,540 came from commercial sources, while the million still under order were being held up by a federal law which required packaging in single-dose vials. The drug companies had only fifty-dose containers. Special permission was obtained for the city to use the larger containers and more vaccine was promised the next day. Another five million doses were ordered.

Police had to use bullhorns and loudspeakers to urge the thousands waiting in lines to go home and come back the next day.

On the 16th, moreover, there was a scare at LaGuardia Airport when a Pan American World Airways plane with nineteen passengers and nine crewmen was quarantined for three hours. A year-old girl on the plane, Judith Vroom, was sick, and since she was coming from Karachi, which was then in the throes of a smallpox epidemic, the disease was suspected. The fears proved unfounded.

A quarter of a million doses were obtained by the city that day and vaccination went on in the city's schools by doctors hired for $8 per three-hour session.

On the 17th something of a world record was set. The city vaccinated half a million people, bringing the total number of New Yorkers to two million. Police stations were doing the bulk of the free work and many reported lines around the block. Half the vaccinations were being done for pay by private physicians, the city reported, and they were now running low on supplies. The city asked more doctors to volunteer for the free program, but out of the thirteen thousand only some four hundred stepped forward. The others were busy charging patients their regular fees.

A man died of smallpox in Camden, New Jersey, across the Delaware River from Philadelphia, and about a hundred miles from New York, but authorities were unable to connect him to the New York outbreak. Vaccination fever now spread to Philadelphia, Trenton and southern New Jersey.

There was one arrest in the city. Sylvia Steinberg, who dressed like a nurse, "vaccinated" five hundred people with water in a Harlem bar to impress her date. She was committed for psychiatric observation.

On the 18th a seaman from New York was taken off an army transport ship in Bremerhaven with smallpox. Authorities could not connect him either to the New York chain of infection.

On the 20th four hundred teams of nurses and doctors hit the city's schools. Every high school student in New York previously unreached was vaccinated that day. The total vaccinated in New York was then 3,450,000. President Truman, on his way to the city on a political trip, was vaccinated before he left Washington.

On the 23rd the Westchester County health commissioner, Dr. William A. Holla, had another complaint: the area's doctors were refusing to help his free-vaccine program because they were more interested in collecting fees. When the Port Chester school physician refused to vaccinate on the grounds that it was not part of his routine duties, local health officers whisked six hundred kids out of the building, lined them up, and vaccinated them while the physician glowered. Holla reported that many other doctors were angry at his free clinics because they were "cutting their business."

But the scare was over. There had not been a case in ten days. The public finally calmed down and the vaccination centers were closed as crowds diminished. Five million people had been vaccinated in a little more than two weeks, probably the largest emergency mass-vaccination campaign in history.

Ibrahim Hoti had reached an exalted position. He had just returned from a pilgrimage to Mecca, fulfilling a goal all Moslems share. He was now a *hajji,* one who has worshiped at the shrine of the Prophet. Relatives from all over came to visit him and share his joy.

Hoti was an ethnic Albanian who lived in Yugoslavia in the Kosovo area, which nestles up to the border with Albania. It is a wildly mountainous region, primitive to an extent found almost nowhere else in Europe. The main overt business was farming and herding. The main covert business was smuggling, and it was believed that Hoti was not above slipping through mountain passes and across the border to bring illicit goods back and forth. He was known to frequent a black market cafe in Dakovica.

But in February, 1972, Hoti was a celebrity. All his Kosovo relatives came to hear of his travels. He undoubtedly told them of his visits to a dervish temple in Iraq and to the market in Baghdad. Probably he did not know that there was smallpox in Baghdad. He showed his purchases to his relatives, and they were grateful and impressed that their family had gained a *hajji*.

Hoti made his pilgrimage the hard way. The Yugoslav government tried to channel the twenty-four hundred or so pilgrims into chartered planes. They were less afraid of importing smallpox than cholera. With the chartered planes the government provided one doctor for every hundred pilgrims, saw to it that they took two grams of tetracycline orally on their return and kept watch on them for five days.

Unfortunately, air flights are an expensive way to become a *hajji*. Hoti and a group of fellow Muslims followed the far cheaper method of chartering a bus. When he returned to Yugoslavia by the bus on February 15, he carried a valid international certificate stating that he had been vaccinated against smallpox on December 19, 1971, in preparation for the trip. It was not stated whether the vaccination took. He had received two cholera shots also, standard procedure despite the fact that cholera vaccine isn't very effective.

The visits from relatives went on for two weeks before tapering off. During that time Hoti saw a physician, as required by law, to present a stool specimen for a cholera

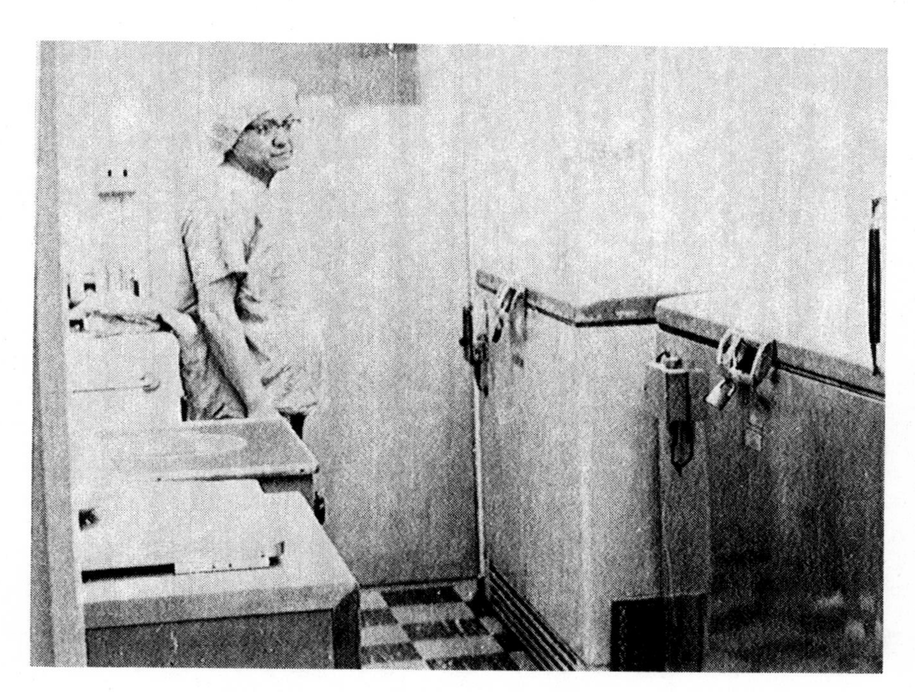

Dr. James Nakano standing by the padlocked freezers at the Center for Disease Control in Alanta, which contain enough smallpox viruses to wipe out the city. *(Joel N. Shurkin)*

Dr. William Foege, with the beard he wore during the smallpox program. *(Center for Disease Control)*

Dr. Nicole Grasset in Geneva. *(Joel N. Shurkin)*

Dr. Isao Arita of the World Health Organization, Geneva. *(Joel N. Shurkin)*

Dr. Victor Ladnyi, at WHO, Geneva. *(Joel N. Shurkin)*

Dr. Larry Brilliant and Girija at their Michigan home. *(Joel N. Shurkin)*

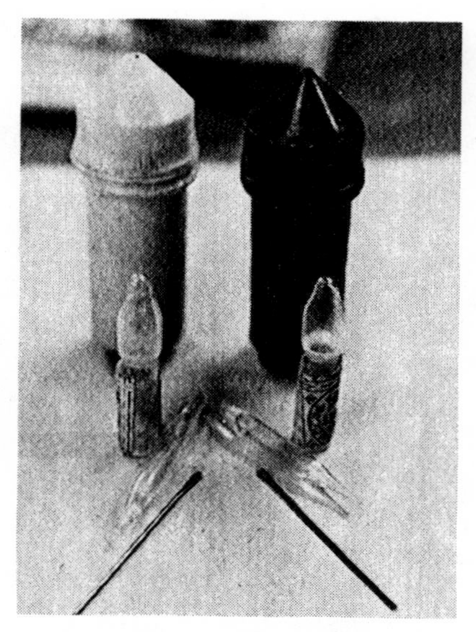

Vaccinator's kit, containing
ꞁwo plastic needle holders, two
ꞁials of freeze-dried vaccine, two
ꞁials of wetting solution,
ꞁifurcated needles. *(Joel N. Shurkin)*

Dr. M.I.D. Sharma outside his Delhi home. *(Joel N. Shurkin)*

R. S. Bajpai (left) and his PMA. R. P. Singhal in Lucknow. *(Martha Brannigan)*

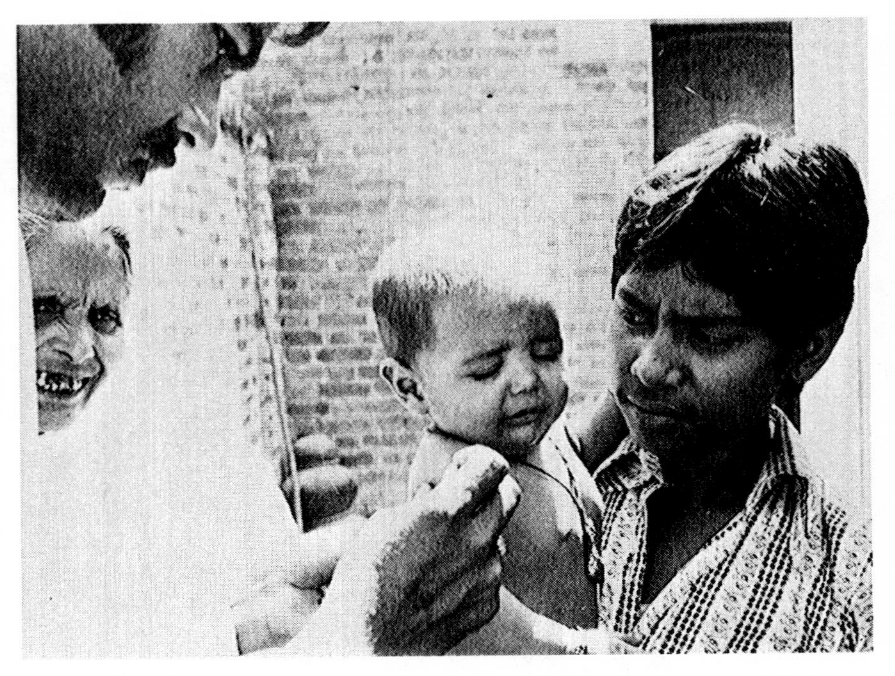

Vaccinator at work in northern Indian state of Uttar Pradesh. *(Joel N. Shurkin)*

Zdeno Jezek checks scars on the face of Ali Maow Maalin, the last known human to suffer from naturally transmitted smallpox. *(Joel N. Shurkin/Philadelphia* Inquirer)

test. The doctor was in Orahovac, the commune nearest Hoti's town. He also visited friends in the Dakovica tavern on the 21st.

To the best of anyone's knowledge, Ibrahim Hoti was not ill a single day following his return to Yugoslavia. He bears no scars of variola.

Everything about Hoti must be taken with a grain of salt, experts agree. He is, as one report stated, "an unreliable historian." Living in the underground world of smugglers does that to people.

It is, nonetheless, a mystery whether he had contracted smallpox. Certainly he did not have a clinical case. Although he showed high titer reactions when tested in April, none of his immediate relatives, who were in close, almost constant contact with him, became sick.

But somebody in that area certainly had the disease. Europe was about to experience its biggest smallpox epidemic since World War II. It would be the first outbreak of any size in Yugoslavia since 1930. It would terrify a nation and send off alarms around the world. It was the best proof possible that so long as the virus lurked anywhere no one in the age of international transportation was completely safe.

As in many places in Europe, protection from vaccine was spotty at best in Yugoslavia. According to an official document, vaccination was compulsory but not always performed. This was particularly true in the Kosovo area, with its poor roads and public services. The ethnic Albanians really have never acquiesced in the fact that they live in Yugoslavia. Few of their countrymen speak their language. They are at the bottom of the economic scale, and they resent both their position and the central government's imposition on their lives.

The government in Belgrade, sensitive to the fact that the country's population is highly diverse, has granted to the six Yugoslav republics and the two autonomous areas considerable independence to run their own affairs, including health services. (Kosovo is part of the Kosovo-Metohija autonomous oblast.) On the day-to-day basis this independence worked fine. When a national emergency broke out, it proved to be disastrous.

Kosovo was perfect for an epidemic in other ways as well. Most of the families are large, with many children, and are crowded into small houses with relatives constantly coming and going. Because of the poor economy many males travel abroad to find work.

And, like most countries in Europe, Yugoslavia had no recent experience with smallpox, so when the disease broke out in the winter of 1972, the government simply did not know the best way to react.

There were eleven cases in the first wave, all of whom were traced to Hoti in one way or another. He had met all six of the patients in Danjane, his home village. There was one patient from the Orahovac commune, a relative who had visited him. At the market in Dakovica he had the chance to infect four others, although nobody knows exactly how.

One of those four, Ljatif Muzza, was the principal vector outside of Kosovo. He was a teacher from the village of Dobri Dub near Novi Pazar and came to Dakovica on February 21 to register for college there. It is not known how or when he met Hoti. On March 3 he became ill and was treated for fever by physicians in Novi Pazar. He finally broke out in a rash and was sent by bus with his brother to a hospital in Cacak. He spent a day there and then was ordered transferred to the Belgrade skin and venereal disease hospital in an ambulance with a pregnant woman (miraculously, she never came down with smallpox). He began to bleed under the skin, and the puzzled doctors transferred him again, this time to the surgery hospital. On March 10 he died. The attending physician blamed internal hemorrhages. In fact, his undiagnosed affliction had been confluent smallpox with internal bleeding. There was no autopsy and the body was shipped back to Novi Pazar for a religious burial. Muzza had enough time to do a lot of infecting.

His brother and one person on the bus between Novi Pazar and Cacak came down with smallpox. In the Cacak hospital he had infected nine persons—eight patients and a nurse. The nurse died. There was an outbreak in Morovic spread by a schoolchild who had been treated for scabies in the skin and venereal disease ward in Cacak while Muzza was there. The child's mother died of smallpox.

In Belgrade Muzza directly infected nineteen patients at the hospital and seven health workers who, in turn, infected five others. Six in this chain of infection died. A person from Trnjane who was in the Belgrade surgery hospital came down with the disease but did not pass it on.

In all Muzza infected thirty-eight people, directly or secondarily, doubling the old European record, in part, at least, because no one diagnosed his case correctly.

By mid-March the Belgrade government knew it had a full-blown epidemic on its hands. Vaccinations and quarantines of possible contacts, particularly Muzza's, began. By this time the second ring of cases was erupting in Kosovo. More than a dozen villages were affected.

Because the epidemic was not really recognized until as late as March 16, many more caught the disease. Yugoslav government epidemiologists traced cases of infection to the hospitals themselves. It was determined by them that of the five other patients in the room with Muzza in the dermatologic hospital in Belgrade four came down with smallpox. In the hospital where he died, he was placed in the intensive-care ward with sixteen patients. Eight came down with the disease. A ninth, who occupied Muzza's bed after his death, and a baby who was merely *carried through* his room also caught smallpox.

Vaccination started in Kosovo on March 16. That same day vaccinators rushed to other foci in the country as well and mass vaccination was ordered. It was immediately discovered that the available vaccine had lost its potency and most of the vaccinations had to be repeated.

Although the laboratory in Zagreb had produced about a million doses of high-quality vaccine, little of it was in the area of the outbreak. The vaccine produced in 1964 that the local governments relied on did not work very well.

According to a controversial report by an observer from the United States Center for Disease Control (CDC) in Atlanta:

> They had no needles, and vaccinated with pens and styluses. Because of the shortage of vaccinating instruments, vaccinators flamed their pens until they were red hot, dipped them into the vaccine, and then vaccinated. Reports were frequent of vaccinees crying in pain from the heat of the vaccinostyle. This problem was common both in Belgrade and Kosovo, and was witnessed by many official and unofficial observers.

As happens in emergencies, even the experts lost their heads. On March 23 and 24 television and radio interviews of one "specialist" listed all the possible contraindications for the vaccine—all the reasons why certain people should not be given it. Many were simply wrong, others were outrageous. So, vaccinations were not administered to infants under one year of age or children over six who had not been vaccinated before. Pregnant women also were considered unsuitable, although there is no evidence of harm either to the mother or fetus and smallpox in pregnancy is particularly ghastly.

Other "experts" warned against washing for twenty days following the vaccination. The Muslim tradition that couples must wash after sexual intercourse meant that the vaccinated in Muslim areas had either to give up sex or washing. Others were told not to drink alcoholic beverages after vaccination or were warned that vaccine administered in sunlight would not work.

The misinformation echoed around the country. When the government did try to vaccinate pregnant women in Kosovo the Muslims refused to cooperate in what they were sure was a cunning form of genocide.

The efforts of local governments to control the outbreak were chaotic at best. Communes made their own arrangements, sometimes with the help of the army, but without informing even neighboring communes. With no clearinghouse for information, travelers might pass from one jurisdiction to another even if they were prime contacts.

With recognition of the outbreak, the federal epidemiologic commission met once a week instead of the usual three times a year. It was this powerful organization that decided to coordinate efforts into a national plan. "However," one document states, "they did not make decisions which affected daily work schedules, epidemiologic priorities, or the flow of information."

The CDC team that was sent to observe found itself with more information than the government had, information gathered by talking to people and making telephone calls.

Not until mid-April was reliable transportation available. The best teams were made up of soldiers, but not being ethnic Albanians, they were distrusted. There were no ethnic Albanian epidemiologists.

Religious sensitivities also hindered the program. The Muslim community was upset that a *hajji* could be suspected of triggering the outbreak; a report states that "there were rumors that Ibrahim Hoti had been ordered not to admit to any illness."

Because the Albanians lived in very close quarters, twenty

to forty in a crammed family dwelling, the second generation of the disease was very large, more than a hundred cases. Because seven layers of clothing were worn in the cold mountains, it took more than three minutes to vaccinate them.

The government began quarantine measures as soon as the epidemic became apparent. By the time the outbreak petered out, 39,702 people had been placed in quarantine, using hotels, hospitals, villages and individual houses as well as specially constructed centers. The measures did not work very well. In the third generation of cases, fifteen people in quarantine caught smallpox. They had left their part of the hospital to join some of the smallpox patients in watching television.

The government tried two drugs to combat the disease, the previously mentioned antiviral agent n-methylisatin-beta-thiosemicarbazone (or Marboran) and blood serum containing vaccina antibodies (vaccinia immune globulin, or VIG). Of the 237 people given Marboran, five came down with smallpox, or better than 2 percent. In the same area 515 contacts were given nothing, and 28 developed smallpox, or 5.4 percent. Since both groups were vaccinated this test didn't prove anything. The drug was discontinued because of the violent side effects. There is no data to indicate whether VIG worked, but past experience with it had not been very encouraging.

The third generation of the disease ran from March 31 to April 12; twenty-three cases were reported, in Belgrade and Kosovo. Many of the cases broke out one full incubation period from the time the disease was successfully diagnosed in the area, so they represented people missed by the government. Three patients in Novi Pazar were contacts of Muzza's brother and developed smallpox while in quarantine. They had been vaccinated unsuccessfully. A number of the victims had not been vaccinated because of the absurd set of "contraindications" promulgated by the "experts." *

Some were infected by initially unidentified second-generation cases. One second-generation case was a little girl who died undiagnosed on March 25. The man who washed her corpse caught the disease.

The last hospitalization was on April 12, and with that the Yugoslav outbreak ended. But there were a number of lessons to be learned and the team from CDC was quick to point them out to their superiors in Atlanta.

The principal factor, the observers concluded, was that the obvious confusion within the public health community in Yugoslavia clearly made matters worse. In November of the epidemic year the Yugoslav government held a national conference on the outbreak. The CDC report by Donald Francis notes that:

> One single problem was evident from the proceedings of the conference, the comments from the conferees and from our experience during the epidemic: the lack of an expert, centralized policy-making center for epidemic control. Without experts at a high central level during the disease outbreak, it is difficult to coordinate any outbreak control. In this conference everyone had their own ideas as to the benefits from quarantine, Marboran, VIG, vaccination, etc., etc., and there was no expert referral leader to answer those questions. This is typical of what happened during the epidemic where individuals without sufficient knowledge of smallpox took it upon themselves to make policy decisions regarding control of the epidemic. This, I feel, was the

* Some experts doubt the United States would have done much better in the panic that follows a smallpox outbreak.

primary administrative difficulty during the smallpox epidemic in Yugoslavia.

Francis states that this probably would not happen in the United States because of CDC.

In total, there were 174 cases and 35 deaths in three generations of cases. It was a lesson for all and provided even stronger reasons for finally eradicating the disease, which had become the target of an international effort in the 1960s.

X

The First God to Fall

In the spring of 1960 a tall, young Ohioan, Donald Ainslie Henderson, returned to the quiet campus of the U.S. Public Health Service's Communicable Disease Center in Atlanta after a year's study at Johns Hopkins. Henderson was to take over CDC's Epidemic Intelligence Service, the government's crack medical detective unit which tracks down outbreaks of infectious, contagious and potentially dangerous diseases in the United States.

EIS worries about such matters as new strains of influenza and about scarlet fever, typhus, malaria, bubonic plague and the various types of food poisoning. One thing it did not worry very much about was smallpox. There had not been a case reported in more than a decade and few people at CDC had ever seen one. Smallpox was not a pressing problem.

Henderson and CDC were, in many ways, ideally

matched. The institution attracts the idealistic, the adventurous, the iconoclastic elite of American medical schools—young men and women who have gone through the conventional medical education and survived with the kind of altruism and zeal that is often twisted out of their colleagues.

"I don't think we're a peculiar breed," one CDC epidemiologist says. "I think it's the system that somehow distorts many people going through it." If most medical school graduates could be presented with the proper challenges, he says, there would be "a lot of takers . . . I think it is an accident of exposure."

Jeans and work clothes are as common as ties and jackets at CDC. Beards and Volkswagens are as common as clean-shaven cheeks and Chevvys. A professional informality and a respect for results seem to dominate the halls of the gray and pink buildings. It is a rare example in government of a bureaucracy that appears to have been kept under control so that it supports the work rather than interferes with it. CDC has earned worldwide respect for its epidemiological work and for the excellence of its laboratories.

Henderson came to it, at least in part, out of idealism.

"He doesn't come on with much of a spiritual nature about him," one colleague says. "He was raised a Calvinist, and a very thick Calvinist streak runs through him. The Calvinist way of being spiritual is to be practical and to do good deeds, and for D.A. [Henderson] it's a translation of what really is that same spiritual training that others have demonstrated in another way. Touch his spiritual button and he'll turn out a smallpox-eradication program."

Henderson, who was not thinking of a smallpox-eradication program when he returned to CDC, has a simpler explanation for his presence there.

"Most of those, including myself, who joined CDC were really not particularly interested in public health. It was two years to serve a doctor's military draft option."

Whatever his motives, D.A. returned with several ideas, including that of getting CDC involved in more international programs. The center really had never paid much attention to occurrences beyond American shores unless an infection of foreign origin threatened to enter the United States. Henderson's experience with new recruits at CDC had convinced him that those coming to Atlanta with an overseas background had a more positive attitude toward public health in general. It was such people whom he wanted in EIS. He was determined to give his staff as much foreign experience as possible. Getting involved with smallpox was not exactly what he had in mind.

Professor Henry Kempe of the University of Colorado was going around the country that year advocating that the United States stop smallpox vaccinations. He said that he had seen no smallpox since the 1940s but too many cases of severe complications in reaction to vaccinations. The vaccinations seemed to him to be a greater danger than smallpox. Kempe was running a referral center and hence saw more cases than a normal physician would see.

Kempe's critics countered that he was exaggerating the problem and urged that vaccinations be continued as a mandatory health procedure. The dispute was fueled partly by the fact there was very little hard evidence to support either side. Some studies in Europe, mainly in Germany and Austria, had shown that reactions to vaccination were not uncommon, but the data seemed to point to a far less serious problem than was evident to anyone in the United States except Kempe.

As part of its legal mandate CDC decided to investigate whether Kempe's fears were warranted. Henderson formed a four-man smallpox team for the purpose. It obtained Department of Labor statistics on the number of smallpox vaccinations performed in the preceding year and then surveyed physicians in four states to learn how many adverse reactions they had seen. The study showed that the problem was not serious in the United States.

The study demonstrated, however, that CDC lacked expertise in smallpox. The U.S. was spending vast sums keeping Public Health Service officers on guard at ports and airports to prevent a single case of smallpox (besides other diseases) from entering the country. But if a smallpox case got through, CDC would not have the experience to handle a possible epidemic. Henderson's team proceeded to broaden its experience. Every time an outbreak occurred in Europe (always by importation from Asia or Africa because it was no longer endemic in Europe), Henderson dispatched someone to watch and get an impression of the disease. The unit's responsibilities were also broadened in the U.S. The unit responded to every report of a suspected case of smallpox in someone returning from an endemic country.

"We did everything from consulting on the phone with the patient's physician to actually picking up a kit, jumping on an airplane, and coming on the scene and bringing back specimens," says Ralph (Rafe) Henderson,* one of the team members.

They got some real practice. In 1962 a Canadian boy traveling from Brazil to Toronto passed through an airport in New York City. He came down with smallpox in Canada. CDC investigated but could find no infections resulting in the United States. The next year a Ghanaian woman developed chickenpox in Washington; a number of physicians suspected smallpox. CDC investigators reached proper diagnosis, but the test laboratory prepared its report incorrectly and some confusion resulted. No one caught smallpox, of course, but the lack of expertise showed again.

* No relation to D. A. Henderson.

Advances made in the technology used to fight smallpox facilitated CDC's involvement.

In the early 1950s Leslie Collier of England's Lister Institute developed a commercially practicable system of freeze-drying the vaccine. Jenner had observed that dry vaccine lasted longer than the liquid form but he did not know how to achieve drying dependably. A method suggested in 1909 was generally ignored.

Collier's technique was to freeze the vaccine quickly in a vacuum. Reconstitution was accomplished with a second ampul of glycerol solution. The World Health Organization's specifications called for a vaccine that remained potent up to four weeks at body temperature. Collier's vaccine could last two years at that temperature.

But administering Collier's reconstituted vaccine was a problem. In the U.S. and Europe the method was to put a drop on a sterilized portion of skin and then scratch it into the skin with a needle. For mass vaccination programs this method did not work very well. In Asia the rotary lancet, a small round instrument with several sharp burrs, was used instead. The vaccine would be placed on the arm and literally ground into the skin. The procedure was very painful and caused terrible scars and reactions. It was difficult to get anyone to sit still long enough for vaccination.

The U.S. Army Research and Development Command thought that it had an answer: its hydraulic jet injector with a special nozzle that could direct vaccine just below the surface of the skin—the right amount and the right place for smallpox vaccine. CDC was given a contract to test the injection technique.

CDC did some testing locally but decided that it really needed a virgin population for a proper trial. It picked Tonga in the South Pacific, a group of islands where there had been no vaccination. Four people—Ron Roberto, Pierce

Gardner, Peter Greenwald and Bill Foege—were sent from Atlanta to perform the vaccinations, study take rates and conduct serologies to test the efficiency of the device. Their experience would be crucial later.

The Tonga team concluded that the jet injector worked quite well and, under ideal conditions, was capable of vaccinating a thousand people an hour.

There was more vaccine testing to be done in the early '60s, tests on a new, more stable measles vaccine. The U.S. government wanted it tested in the field on a mass scale. The Bureau of Biologics chose the inland African country of Upper Volta for the test. Measles vaccinations by jet injector were given to more than ten thousand Voltaic children in the summer of 1963.

The government of Upper Volta was very pleased at the number immunized and wanted all their children vaccinated against measles, a disease far more serious in Africa (where it claimed one child of every 20 who contracted it) than it is in Western nations. Other African countries also wanted measles-vaccination programs. The U.S. decided to send two people from the bureau to Africa for one year to train vaccination teams from each country applying in the use of the jet injector. The program was under the auspices of the U.S. Agency for International Development (AID), the main foreign aid arm of the government.

AID wanted some help from CDC.

"I got a call from AID asking, 'Could you send somebody out, just to be our eyes and ears, to see how things go?' " Henderson recalled. "We said, 'Fine, for six weeks.' " Henderson chose Lawrence K. Altman,* an epidemiologist with EIS. It was CDC's chance to get into more of international medicine.

---

* Dr. Altman is now a medical writer for the *New York Times*.

"Larry Altman went out for six weeks, which went on and on," Henderson said, "and became more like six months because the whole thing was a bloody debacle. The vehicles were huge and did only four miles per gallon; they were forever getting stuck in the sand . . . magnificent, great big bloody things.

"The vehicles had a small butane flame that was to run the refrigerator. There was a water supply system, so you could wash your hands. You had everything on the vehicle. But the flame used for the refrigerator was right beside the gasoline tank. There were repeated episodes of these trucks going up in smoke. The refrigeration didn't work all that well either.

"There was a notable exchange between Larry and AID people when Larry complained that the refrigerator wasn't keeping the measles vaccine cold. They said, 'Well, park it under a tree, keep it cold.' And he fired back something like, 'Please send a hundred Dutch elms. I'm in the middle of a treeless desert!' "

It was only the first exchange in what would soon become a running series of battles between CDC and AID.

The next year Henderson received another telephone call from AID.

"They appreciated that things hadn't gone as well as they had hoped, they now wanted to give measles vaccine to all the French-speaking countries of Central and West Africa and suggested that it would be nice if we could assign nine of our EIS medical officers there for six months to help with this program. We just did not have all that many staff. To take some of them in May, when all the assignments were set, to pull out nine guys and send them for six months, was impossible," Henderson says.

"At the same time I hated to say no to AID, because we really wanted to work with them, to post various consultative sections with AID overseas." But the measles program bothered Henderson. It was not a one-shot deal. The deadly time for the disease is between six months and three years of age. If CDC vaccinated everyone of that age in the area they would have to do the same thing all over again in a few years as a new generation of susceptibles was born. "How long are you going to keep this up?" he asked. The countries could not carry on without permanent assistance.

AID didn't know. It was decided to try to convince AID into changing its mind without harming future relations between the two federal agencies. CDC decided to make AID an offer it couldn't accept. CDC's negotiators decided to propose linking the measles-control program with a smallpox-eradication program in a contiguous block of eighteen African nations. One of those countries was Nigeria, a former British colony.

There was a firm AID policy at the time that no health program would be conducted in Nigeria, one of the potential giants of the continent, because AID felt that economic development was needed first. A health system would be a natural result, AID felt. Moreover, Nigeria's estimated sixty million people made a mass-vaccination program very difficult.

"So we proposed what was patently unacceptable to AID as a matter of policy. We would be happy to undertake both smallpox eradication and measles control throughout this block of countries. . . . We knew this wouldn't fly," Henderson says.

That was in May, 1965. Henderson was so sure that AID would retreat that he agreed to send Don Millar to London for a year of study, telling him, "Nothing is going to happen on the smallpox front this year."

"AID sat, thought, scratched its head, worried, fretted, stewed," Henderson says. "We developed some estimates of cost, and so forth, and figured nothing would happen."

CDC couldn't get an answer from AID.

The decision was taken out of everyone's hands. A minister of one of the African countries—no one remembers exactly who—met with President Lyndon Johnson and praised the measles program as the kind which the United States should be doing more of. Johnson, pleased to hear praise of his foreign policy for a change, passed the word to AID: Do more vaccinating. AID went to the CDC and, to the astonishment of all concerned, agreed to Henderson's plan.

"Was my boss, Alex Langmuir, exactly happy that this had come about?" Henderson muses. "Happy? Christ, he was furious!"

Langmuir was not alone. Linking a smallpox program to the measles-control efforts disturbed many. Dr. Benjamin Blood, then associate director for international organization affairs of the Office of International Health, said that many were reluctant to accept the plan, even with the White House pressure.

"There was some hesitancy, in fact quite a lot of concern, that this would dilute the measles program which had been given such a high priority," Blood said. "It came mostly from AID—not so much from the technical people. There were others who didn't truly understand." Blood * was to play the reluctant role of referee in the ensuing battles.

"AID tried to persuade us to make it a smallpox-measles eradication program," Henderson said, "and we kept saying, 'No, we really can't guarantee measles eradication. We don't understand the disease that well. We haven't eradicated it in the United States and we aren't about to promise more than we can deliver.' Measles spreads more rapidly than smallpox, the vaccine is not so stable. . . . Whereas we felt that we

* Blood is now head of primate research at the National Institutes of Health.

could eradicate smallpox, we just didn't know about measles.

"But, lo and behold, we got on the plane in November, 1965, and went to Africa."

Henderson was accompanied by Warren Winklestein and CDC's Henry Gelfand. Both knew a little about Africa and Winklestein spoke French. The CDC people attended the ministers' meeting held in Bobo-Dioulasso, Upper Volta, then went on to Nigeria. They touched base with the AID mission there ("which was distinctly unhappy with this whole prospect," Henderson says) and continued to the Cameroons to meet with more ministers.

On the way back to the U.S. Henderson dropped in on the World Health Organization to get some money for gasoline. He had found that AID refused to pay for fuel.

"This policy was perfectly evident," he says. "Many of the programs were doing nothing because although they had the vehicles, the personnel, the equipment, the supplies, they had no money for petrol. AID said that the countries (getting assistance) should provide this. But to the (local) countries, this was foreign exchange, and they were really short of foreign exchange. They couldn't provide the petrol costs. So AID had marvelous plans but the vehicles didn't move. . . . Everybody shrugged their shoulders and nothing happened."

WHO had been nominally committed to eradicating smallpox on a global scale since 1958 but was actually doing very little. According to Blood, the U.S. was trying deliberately to link its bilateral West African program to WHO as a sign of good faith and to encourage more WHO effort and demonstrate that eradication was possible.

"WHO was very interested," Blood said, "and more and more when it saw that we were not only talking but were putting up. There was no way WHO could ignore our program. On the contrary, it had a mandate to be supportive everywhere."

So WHO was supportive and Henderson got his gasoline.

Back in Atlanta Henderson and the smallpox unit were trying to get organized. They needed a staff, a manual, separate plans for each country and agreements with the governments involved.

In the eastern Nigerian area called Yache, Bill Foege, who had been part of the Tonga team, was now serving as the medical officer in a mission run by the Lutheran Church, Missouri Synod, for native Lutherans. Foege had left CDC in 1964 for missionary work. He is a quiet, active man, who stands six feet seven inches tall. He is described by one of his compatriots as "one of the most profoundly Christian persons I have ever met."

Foege was in Nigeria trying to see if medical missionaries could make preventive medicine work under the primitive conditions frequently encountered.

The mission consisted of twenty families; Foege's wife and two children were along. It was linked to other missionaries in the area by a radio network. The mission was devoted to medical assistance and to linguistic study. There were twenty-five languages in Yache, divided into twelve distinct groups.

During the summer of 1966 Foege received a letter from Henderson telling him about the smallpox-measles program and asking for Foege's help. Would Foege consider being a consultant for about a year to help set up the program?

A reluctant Foege consulted the mission's superior, Dr. Wolfgang Bulle, in St. Louis. He explained that he would not have to leave the mission entirely but hoped to spend weekends there while working for CDC the rest of the time.

Bulle agreed that this would be appropriate. Foege never went back to missionary work.*

Foege went on the CDC payroll and moved ninety miles to Enugu, the capital of the eastern Nigeria region. (It was a moderately complicated financial arrangement, Henderson said, "which was not, I think, entirely kosher within AID rules.")

"We set a target date of January of '67, a little more than a year hence, against all assurances from AID that nobody ever got anything like this underway in short of three years and that realistically we were being stupid," Henderson says. "We went through this in diagram and God knows what-all to show that it could be done. And in fact, we did recruit people, we did get the manuals drawn up, we did get a fair proportion of the supplies out there by January, '67, and, by God, we began a program.

"I must say, it was a helluva year!"

One of the first on the scene was Ralph (Rafe) Henderson. Henderson, holder of four degrees from Harvard, was sent as AID's agent assigned to the health organization of the former French colonies based in Bobo-Dioulasso. As the technical consultant, he worked also with the U.S. mission in the Upper Volta capital, Ouagadougou.

The landlocked Upper Volta is named for the tributaries of the mighty Volta River which flow through its territory. A poor country with almost no natural resources, most of its six million people follow traditional religions or Islam. Upper Volta comprises mostly woodland and pasture, except in the north, which is severe desert. The country had been independent for six years when Henderson arrived.

---

* Foege is now director of CDC, which is now called the Center for Disease Control.

"I had never been to Africa before," he says. "I couldn't do anything wrong because I didn't know enough to do anything. Whatever I succeeded in was a step up. I didn't go with expectations of being able to do much.

"What it amounted to was going around and talking to ministries of health, looking through their eyes at the problems . . . and trying to figure out if there were any solutions that one could suggest, not on an individual but on a systems basis. What kinds of things could we set up in the future?"

The problems became immediately apparent with the arrival of the first equipment.

"The trucks, the vaccine, the jet injectors were sent over," Rafe Henderson recalls, "and in theory I was there to make all those things work within the local system." But soon spare parts were running low. Once even the glycerol solution to reconstitute the vaccine ran out. Bottles of Evian water seemed to work just as well.

"I got a very good grounding in what could go wrong with an immunization program in those six months."

There were, for instance, coups d'état. Henderson and the political officer from the U.S. embassy were in Ouagadougou when the army revolted and besieged President Maurice Yaméogo in his palace. The two Americans were caught in the siege. They explained to the soldiers that they were foreigners with the immunization program who merely happened to be wandering through. The soldiers, apparently impressed, opened their lines so that the two men could get away. Then they resumed firing at the palace and completed Yaméogo's overthrow.

Rafe Henderson and Foege returned to Atlanta to train the forty or so CDC people D. A. Henderson had recruited. The staff had been drawn from the main Atlanta complex and from state health departments around the country.

In October Rafe Henderson and Foege went again to Africa. Foege had a new child, Henderson a new wife, and the trip was not without trepidation.

"I was very concerned," Henderson says. "My wife was a displaced person during World War II. She's Latvian. . . . I said to myself, 'Just getting married, I'm going to take my wife back to a very uncertain situation. Isn't this kind of risky?' It turned out that she loved it."

At the time West Africa had one of the highest rates of smallpox in the world. The exact number of cases is not known because the reporting system was so poor, a problem that had to be solved quickly. The total reported to WHO in the world in 1967 was 131,418 cases. There were about that many cases found in northern Nigeria alone!

The prevalent intermediate strain was particularly deadly to children, killing 20 to 30 percent of those struck. Among adults the death rate was 8 to 10 percent.

"We would see cases in their late stages," Rafe Henderson says. "To see a bad case, the reaction is one of horror, of wanting to withdraw. The person is just covered with blood and scabs and the eyes are closed and puffy. Breathing is hard. Your fear is not only that you can do nothing for him but your reaction is, 'Oh, my God! Could I get this?' "

In general the countries of West Africa had functioning health systems, including mobile teams that roamed the bush providing basic services, including vaccinations. Most African health officials doubted that smallpox could be eradicated. If it hadn't been done during the colonial period, a bunch of Americans running around with jet injectors wouldn't manage the job. The first task was to conquer the local doubts and train the health teams.

Health teams had been used in an earlier program to eradicate yaws (a contagious nonvenereal disease with characteristics similar to syphilis) and knew how to operate in the field, but they had to be trained to use the jet injector and the stable, potent smallpox vaccine.

"It took us a while to get our supplies," says Foege. "We were in Nigeria in the fall of 1966, but really didn't get supplied until January of '67, that was one of the things we were facing when we went to the first smallpox outbreak."

On December 4, 1966, Hector Ottermuller, who was near Foege's old mission, radioed that he had a smallpox problem. The village involved was seven miles from the road, and Foege and Ottermuller had to make the last part of the trip on motorbikes.

"We were faced with the dilemma of how to control it, because we really didn't have the supplies that we would want, and our preconceived strategy had been to really blitz an area with vaccine in a case like this," Foege said. "So we were faced with a question: how do we do the most with a few supplies?"

First Foege set up a surveillance network making use of the missionary radio system. Regularly every afternoon the missionaries of eastern Nigeria radioed one another. So on his first afternoon Foege took to the air and asked them to check for smallpox. He assigned each missionary to a specific area and asked for radio reports the next afternoon. Thus he learned of several more outbreaks.

The next problem was to contain them. In the affected villages Foege and his aides vaccinated everyone that they could find and then tried to predict where the disease was likely to spread.

"We tried to take on the mental characteristics of a smallpox virus trying to achieve immortality," Foege says. "Where would you go?" It was decided that the virus would follow the routes taken by people in marketing and that at markets there would be the main concentrations of human hosts. Foege vaccinated in the three likeliest places. The strategy proved sound, because in two of the three the disease later was found to be incubating.

Not all the important lessons to be learned from the Yache outbreaks became clear until a few years later.

That more help was needed was immediately obvious. CDC's first thirty-seven recruits (fifteen medical officers and twenty-two operations officers) were soon on their way. For many it was one of the greatest adventures of their lives, something they would never forget. For others it was unmitigated horror. A few got off the plane, looked around, and flew right back to the United States. Some of the men were pressured to return by the unhappy wives who accompanied them, and not a few marriages eventually broke up as a result of the tension created by Africa.

The wives seemed particularly susceptible if their husbands were based in the field rather than in the capitals of the nations to which the campaign eventually extended. The family of the smallpox fighter in the field suddenly was forced to be self-contained, supply its own entertainment, provide its own resources, spend all of its time together. That is a heavy weight to put on some marriages, because in constant close contact, without outside distractions, there are families whose members find that they don't really like one another. Wives who had had jobs back home were frequently unhappy because they could not find meaningful work in Africa. For such women spending all their time with their children and in enforced idleness might prove a strain.

They were the minority, however. Rafe Henderson's wife became his assistant, traveling across Upper Volta with him and becoming a first-rate field epidemiologist. A few other women did likewise. Others found outside interests to distract them from the boredom. One woman decided that she was not going to pass up the opportunity to see Africa. So one day she waved good-bye to her husband and children, climbed aboard a "mammy wagon," a native bus, and disappeared for three months. She traveled three thousand miles in buses and trains by herself and finally returned with no more ceremony than a "Hi," as if she had only gone to market.

Not all the stations were unpleasant either. George Stroh, who was based in Nigeria for a few years, found himself in Jos, far removed from what is conventionally considered civilization. A town of about 120,000 people situated high on a plateau with delightful weather, it had been used as a rest area for British troops during the North African campaign of World War II. Several hydroelectric dams provided power for the local mining industry and for private use. Stroh's children went to a good mission school and were perfectly safe walking the streets. Discussing it years later, Stroh remembers Jos as "one of the nicest places you could possibly live in."

An additional factor in making life pleasant was that the CDC people were now working for AID, and AID takes care of its own—to a fault. "We were provided housing by the Nigerian government," Stroh says, "but when you moved in, AID would pay a contractor to screen the house and install a freezer, refrigerator, washing machine, stove, air conditioners. There were no problems at all. Everyone had a vehicle. . . . It cost me less money for gasoline in Africa than in the States. The standard of living for most American AID employees was better than their standard of living in the States. You had servants running all over the house."

Stroh said that it took a while to get used to the fact that you were not expected to get your hands dirty.

It took no time at all, however, for relations between CDC and AID to fall apart. The two agencies were fated not to get along.

CDC people are naturally iconoclastic, individualistic and

antibureaucratic. "AID often perceived the CDC staff as an extra burden," Rafe Henderson says, "and the CDC staff wasn't always good in saying, 'Yes, we're working with the whole mission system.' It was more like, 'We're from CDC and we're really going to do this job the way nobody else has before.' In some instances it certainly generated a feeling of competition and ill will that, I think, was more and more manifest as the smallpox program showed increasing success."

There were roaring fights, not only in Africa but back in Washington also.

Says Ben Blood, who was usually caught in the middle, "It was quite clear in CDC's mind what had to be done, but AID had a system of organizing, planning and operating programs that didn't necessarily follow the CDC pattern. So basically, there was an awful lot of accommodating necessary. And at times it looked as if it was going to fall apart totally."

But it did not fall apart completely. The CDC epidemiologists and national public health workers quickly set up an efficient system to make sure that as many people as possible were vaccinated. The local staff did the vaccinating; the operations officers from CDC in the national and regional capitals spent much of their time in the field training, checking results and assisting where needed.

Some slept on cots in the bush or under a clump of mango trees. Sometimes village chiefs would turn over a hut to the visitors and stock it with live chickens for dinner. Usually there were too many chickens, and the clucking and pecking kept everyone in the hut up all night.

It was never clear what the people vaccinated understood. What assured the vaccinators and the CDC workers a warm welcome was that most of the villages were never visited by foreigners, rarely saw health workers and associated a great deal of prestige with arrival in a big truck. The Americans enjoyed the advantage that the villagers knew that they had controlled the dreaded measles epidemics and with one shot of penicillin had cured the terrible affliction of yaws. Injections and inoculations were considered powerful medicine in West Africa.

Powerful medicine was definitely required in the contiguous areas of Nigeria, Dahomey and Togo (now the People's Republic of Benin), where a smallpox cult centered around the god Sopono was practiced. Sopono was not to be the last god of smallpox that the eradication program would run into but in some ways was the most terrible. The worship was both to prevent smallpox and to assure a safe recovery should it strike. Since the priests of Sopono obtained all the worldly goods of those who died of smallpox, they had a vested interest in its spread. CDC's Stan Foster says that in one outbreak in Dahomey priests were seen collecting scabs of victims. He suspected that they would smear the scabs onto sharp sticks, which would be placed in doorways where the unwary might suffer a scratch and catch the disease.

As a countermeasure one vaccinator put up a morale-boosting sign that translated into "Let's Kill Sopono." Vaccinations continued few until the wording was changed to urge an attack on the disease, not the deity.

The CDC team had to learn that smallpox was less contagious than they had feared. While there were a few cases where the virus wafted through an air duct triggering an outbreak, by and large, that kind of contagion is unusual.

"The first time I saw a smallpox case," Rafe Henderson says, "I got covered from head to toe in a gown and gloves and went in to see the person. . . . My fear wasn't for myself but of spreading the pox virus in the community afterwards. After several cases, the sterile techniques began to break down, and in the end you'd just walk on in, look at the patients, say what you could to them and then walk on out. You'd wash your hands. You'd make a big thing about being in contact and spreading the scabs, but there wasn't any of the superisolation technique that . . . would be absolutely appropriate in the United States.

"But where smallpox is endemic, you're not adding a lot to the virus pool by such casual contacts. We grew very, very casual in our contacts. And even at that, we were a lot more careful in our contacts than some of the local medical authorities, who wouldn't take any precautions at all."

The system developed to control the disease and continue the mass-vaccination program was simple shoe-leather epidemiology. The health workers went from village to village, trying to vaccinate all the children that they could find. To keep a check on how well they were doing, CDC tried to match the number of children they saw with the number the census said were living in a village. This procedure proved useless.

Instead assessment teams followed the vaccinators and used random sampling techniques developed by pollsters to check up. A village would be selected at random. The assessment team would ask the headman what hamlets were dependent on the village. Most of the villages and hamlets were interlocked in a complex system that involved economic, social and familial dependency. If the headman said, for instance, that fifteen hamlets were under his authority, one would be selected at random and visited. Again the hamlet headman was contacted to identify the individuals responsible to him either there or elsewhere. Each person named would be tracked down to see if the vaccination had been administered, and the performance of the vaccinating team then could be evaluated reliably.

Another technique was the market survey. The markets are the social and economic cores for each cluster of villages.

In Upper Volta the saying is that for a market to be a good one, a man must be able to meet his friend, his enemy, his lover and his wife's lover there.

The markets are circular, each village holding a pie-shaped wedge that gives it straight-line access to the market. In case of fights residents of different villages can flee the market without having to cross each other's path.

This setup enabled the assessment teams to use random sampling techniques to examine the people at the market for vaccination scars. Because they could determine where a village lay by the location of its merchants in the circle, they could track down places which the vaccinators had missed.

Still another technique was to visit a village, make a survey and then up to a distance of ten kilometers stop people at random. Nomads or farmers in the fields might be the target. Such people were the most likely to be missed in the mass-vaccination program; if they were covered, the vaccinators would have done a good job.

Such fieldwork, incidentally, merely added to the animosity between CDC and AID, the CDC people felt. "They sort of half jokingly referred to you as a 'bushman,' " Stroh says, "which meant you really weren't quite conveying the image you should convey as an official of U.S. AID." AID also might have been disturbed by the fact that CDC people were proving to be jacks-of-all-trades—out of necessity—and did much of their own work on their own. Spare parts were few and far between and equipment broke down. Sometimes they ran out of money. Stroh and some Nigerian medical students contrived their own cold room for the measles vaccine when the ministry of health said that it couldn't afford to build one. If the jet injector broke down they would visit marketplaces to find adaptable parts. Improvisation was vital.

A moral problem also confronted the epidemiologists:

How were they to limit their work to smallpox and measles vaccination when primary medical care was desperately lacking? They would be faced constantly with people who needed doctors and medicine and could not help.

"We didn't have the supplies and we knew we'd be wiped out," Rafe Henderson explains. "That was one of the arguments early on: Could you go into an area with such medical needs . . . without providing regular care in addition to preventive services? Would the public accept a team that said, 'I'm sorry, we're not here to do anything else but give you this shot in the arm'? It turned out that we were accepted very warmly."

In obvious cases of emergency the trucks would be used to take the sick or injured to the nearest health center or hospital but that was the extent of the primary health care given by the CDC teams.

"If you're interested in the humanitarian goal of preventing suffering from a disease," Rafe Henderson says, "how do you best use your resources? The clear answer is, you stop the transmission of the disease. That does not involve a lot of care for sick people."

The people seemed to understand. In Muslim areas such as northern Nigeria Stan Foster would visit the emir of a district and explain that he wanted to vaccinate everyone within five miles.

"I remember those days very clearly. We would meet with the emir, in his flowing robes, and his district council. We would explain what we wanted and they would set up a plan for a vaccination site. . . . You would get there about 5:30, 6 in the morning, and there would be 6,000 people in line, waiting for vaccination. . . .

"Being a Muslim country, women and men had to be done separately; usually it was the men first, then the women and children. . . . We had some women vaccinators, but this wasn't a problem. In the traditional Muslim towns like Kano we had to vaccinate the women at night. This was a great social occasion for them. They dressed up in their finest."

Sometimes it was too much for the women, few of whom were used to crowds. They would panic and break through doors and windows.

Frequently the emirs would insist on witnessing the vaccinations to add their authority to the proceedings. With such assistance, CDC found 93 to 97 percent coverage.

While Foster saw one team vaccinate 14,000 people in one day and many local governments supported the program and believed in the mass-vaccination techniques, the program did not work very well where local authorities were weak or population density was high. CDC found that smallpox was not being eradicated in all of West Africa.

It would require reinforcements, another war with another reluctant bureaucracy, allies from unlikely places and a complete change in strategy before Sopono really would be the first god of smallpox to fall.

XI

The Reluctant Warriors

The headquarters of the World Health Organization sits atop a lush, manicured hill on the outskirts of Geneva. On a very clear day, when the winds off the Alps whisk away the mist formed over the lake, the workers on the top-floor WHO cafeteria can see Mont Blanc and Chamonix, forty miles to the east.

A half mile below the hill the Palais des Nations, the U.N.'s European headquarters, basks in the sun. Its walks are lined with flowers. The only sound, besides the bees, is the rustle of the breeze on the lindens and oaks.

Geneva is one of the world's most pastoral and beautiful cities, the elbow of an arm of Switzerland that carries Lake Leman into Haute-Savoie. It is a quiet, prosperous town, with prices among the highest in the world. Mysteriously, the Swiss seem to cope. Only the tourist and visitor from outside seem to notice how difficult it is to get a decent meal for under $10, or a hotel room for under $50.

The people who work for WHO and the other U.N. agencies form their own little subculture. They are the "internationals," the civil servants who now owe their livelihood and allegiance to organizations that ignore every frontier. After a few years they acquire their own accent, a kind of mid-Atlantic idiom colored by the accent of their national origin. Their apartments are cluttered with the handicrafts and art of a dozen nations: silk rugs from Thailand, woodwork from Nigeria, silver from Mexico. Their conversation, even when they talk shop, is cluttered with the exotic places and people they meet in the general course of their business. Their workplace is the world.

Some with WHO are veterans of their national health ministries, sent to pass the remainder of their productive years waiting for their tax-free pensions. To these people, anything that ruffles the halcyon waters of their lives is to be avoided. Some are genuinely idealistic, working for WHO because they want to make a contribution to the people of the world. To them the placid atmosphere is probably something of a problem.

It is a peaceful setting, too peaceful, perhaps, for an agency devoted to alleviating the diseases of poverty—directly concerned with the filth of a Dacca slum, the vermin of Brazilian jungles, the crawling things in the slime of the Congo and the dust of the Deccan plateau. One almost wishes that they would plant a child with scabrous yaws on the rich green lawn just to remind those inside the modern glass-and-concrete air-conditioned building why they are there.

They were almost dragged into the smallpox-eradication program. Many of them rose to the occasion and attained heights that idealists should envy. Others, terrified that their world was being disrupted, sank to petulance and obstruction. But the end result would be WHO's finest hour, a time when an international organization made a rare lasting mark on world history and unquestionably earned universal gratitude.

There is some irony in the fact that it all began in one of the rare occasions when the World Health Assembly, WHO's governing board, agreed to meet outside of Geneva. In 1958, at the urging of Senator Hubert H. Humphrey, an ardent supporter of the U.N. and its agencies, the assembly met in Minneapolis, Minnesota.

The Russians had a surprise in Minneapolis: Victor Zhdanov, the Soviet vice minister of health, had a resolution to offer. Zhdanov, an extraverted intellectual who could lace his speeches with quotations from Jefferson with as much ease as he could cite Lenin and Marx, proposed that WHO eradicate smallpox. Befitting a Russian, he proposed a five-year plan: For the first two years all countries would produce as much vaccine as possible, and for the next three a mass-vaccination program would reach everyone in the world who had not been vaccinated.

The Soviets' proposal came against a background in which several previous attempts to eradicate a disease—to render it completely extinct in nature—had failed miserably.

In 1909 Dr. William Crawford Gorgas of the U.S. Army Medical Corps suggested that it was possible to eradicate yellow fever. The method proposed was to destroy the *Aedes aegypti* mosquito in its breeding areas in endemic nations, thus breaking the chain of infection. In 1915 the Rockefeller Foundation's yellow fever commission attempted this method in Brazil and failed.

There was a gap in science's knowledge of yellow fever. The mosquito certainly was the principal vector—path of transmission—in urban areas, but it was not the only way yellow fever could be transmitted. Small jungle animals, usually monkeys, also harbored the virus, and no matter how well the mosquito breeding areas were sprayed the disease always came back if mosquitoes bit the monkeys and then bit humans. And yellow fever is still endemic in many parts of the world although it's fairly uncommon.

More to the point, WHO then was enmeshed in a disastrous malaria-eradication program foisted upon it by the United States. A program begun in 1950, sponsored by the Pan American Health Organization and engineered by Dr. Fred Soper, had wiped out malaria in all but a half dozen Latin American nations. The U.S. believed that if PAHO could be successful, so could WHO.

Here again, the idea was to spray the breeding areas of mosquitoes of the *Anopheles* genus to control and eventually eradicate the malaria-causing parasite that they carry. Combined with chloroquine, the universally accepted drug which effectively controls malaria, it was felt that the disease could be exterminated.

But the eradicators ran into evolution. Some mosquitoes by chance were genetically impervious to DDT and the other modern insecticides; they reproduced quickly to fill the ecological niche left vacant by the susceptible insects. The parasite too evolved, producing a strain immune to chloroquine.

Malaria, the scourge of the tropics, is now a greater threat to the public health in developing nations than before the eradication program began. In 1977 there were an estimated 15 million cases in India alone.

So WHO and its director-general Marcelino Candau were not enthusiastic about Zhdanov's resolution. Malaria eradication had been oversold, and they were afraid the same was about to happen with smallpox eradication.

The Soviets' motives were not clear, at least not to American diplomats. Perhaps the Soviets were just trying to find a program to emulate the ill-fated U.S.-backed malaria effort. Perhaps their motives were simply humane and they thought that smallpox eradication by mass vaccination was feasible and should be done for its own sake. No doubt the

Soviets noticed how many endemic countries touch their border.

Mass vaccination was considered by all to be the proper method for the eventual eradication of smallpox. "Theoretically, it's very easy to understand," says I. D. Ladnyi, a Soviet assistant director-general who was to head the program in part of Africa. "Let us say that if even twenty years ago everybody had been vaccinated at the same time against smallpox, there would be no more transmissions. That's why many people ... did their best by mass vaccination to achieve the interruption of transmission."

In Geneva the next year (1959) the WHO assembly passed the resolution. WHO was now committed to a mass-vaccination program for the worldwide eradication of smallpox.

Trouble began immediately. To implement the program, an Egyptian, Sidky, was appointed the first smallpox medical officer. He asked for contributions from member nations, and some trickled in. Few of the countries would come up with more than $100,000, so the campaign quickly floundered. WHO and the United Nations Children's Fund (UNICEF) taught a number of developing countries to make the vaccine but the quality was not consistent. There were no vehicles, there was no equipment, no manpower and no money.

Sidky left under a cloud. The Soviets began to fume. Every year they would go back to the WHO assembly, hoping in vain that something positive would be reported. A few symposia were called but most produced nothing but meaningless verbiage. Geneva's lack of enthusiasm could not have been clearer.

The few countries that launched programs in conjunction with WHO, most notably India, with almost no exception failed completely. The eradicators, wedded to the mass-vaccination philosophy, simply found that they could not vaccinate everybody. Mainland China did have smallpox eradicated as early as 1960 by literally ordering everyone to be vaccinated. Few countries had the power of the Peking regime to force such compliance.

The failure in most countries was so palpable that many governments thought of abandoning the program completely and just trying to control epidemics.

At about this time D. A. Henderson showed up in Geneva to beg gasoline money for his African smallpox program. Candau was required by the 1958 resolution to be cooperative in these matters but he was not enthusiastic. Although he said later that he regretted his decision, nonetheless he provided the money Henderson requested. WHO thus became inextricably linked to the U.S.'s bilateral programs, which by then encompassed twenty African nations, from Brazzaville-Congo to Mauritania, from Chad to Liberia.

Additional pressure was building on WHO to make a major effort. "The U.S. had been giving serious consideration to what its priorities were," Ben Blood remembers, "and it was not until the middle '60s that a definite policy had been worked out to get behind eradication in a major way."

The Russians also were pushing. Many of the developing nations were pushing. It all came to a head in 1965.

The WHO assembly found itself with a new resolution, fathered by several countries, specifically dedicating WHO to eradicating smallpox in ten years.

"Our position," says Blood, "was to insist upon a resolution to eradicate smallpox within a decade. This was the new feature, that we were more than committed, that we had made eradication the highest priority, and that it be completed in a decade. Certainly it indicated that we were prepared to make a special contribution." To emphasize the point the American delegates read a message from President Johnson urging approval of the resolution.

The proposal went to the executive board of the assembly

in January, 1966. From the debate it was apparent that the resolution would be passed and the director-general was asked to come up with a budget for the program to be presented to the assembly in 1966. Candau was displeased.

He asked one of his experts, Charles Cockburn, to draw up the program. Cockburn and some assistants produced a program with an allotment of $2.5 million within the regular WHO budget, on the assumption that many nations, particularly those strongly backing eradication (chiefly the U.S. and the USSR), would contribute two or three times the amount. WHO would act as the central coordinating group for the expenditures.

According to D. A. Henderson, Candau seized upon the budget as a device to squelch the program. "Let's not say he was against it," Henderson asserts. "He was totally against it! Malaria control was going badly. Candau had worked with the Rockefeller yellow fever commission. He knew Brazil (which was still endemic), he knew the Amazon and what an impossible area it was. He knew that you couldn't get rid of malaria there and couldn't see how you could get rid of smallpox. Frankly, he just didn't understand the epidemiology of smallpox. He's said to me several times since, he still can't figure out how the Central Americans got rid of smallpox. But he didn't want the program!"

Blood is not sure how deep-seated the animosity was. "I was never quite sure how much was real skepticism and how much was just to play the devil's advocate. I know he did certainly express doubts from time to time."

Candau decided to use the budget to his own ends. Every year the director-general submits a proposed budget and kind of balances this off between the developing countries who want the budget to go up more, and the developed countries who don't want it to go up at all, or very little. So he put in the regular increase for the budget and then said, "*On top* of this I'm proposing to put in $2 million more for smallpox." His feeling was that the developed countries would shoot the smallpox program down in flames and that would be the end of it.

But Candau misjudged. Many of the developed nations were spending millions every year in attempts to keep people with smallpox from crossing their borders. The United States spent $150 million that year (1967-68) alone. The developed nations had genuine economic reasons for wanting an eradication program, perhaps more than the endemic countries, where smallpox was just one of many things to worry about.

Action was swift. The assembly met in May, 1966, and the resolution passed with only a few negative votes and a few grumbles about the size of the additional budget. The British, for instance, wanted it cut to $1 million, but the full increase carried.

Sadly, one of the dissenting votes was cast by the United States. Blood says, "Quite frequently in those days, we were first to call for new program initiatives on the one hand and then voting against the budget on the other. The initiatives for the new program largely came from the Public Health Service, as benefiting world health and including, certainly, the health of the American people. And that was the mission of the Public Health Service.

"The State Department had the problem of funding all the international organizations, including WHO. They had the decision, depending on how Congress was acting at the time, of having to cut budgets or not, and it didn't make any difference to them whether it was WHO or some other international organization. The position was developed on that basis, without regard to program interests. It was kind of awkward."

"It was weird," says D. A. Henderson.

Candau was reported to be very unhappy with the results of the vote. He decided that since it was the United States and D. A. Henderson that largely had brought WHO into the smallpox program, the United States and Henderson would have to share the responsibility and blame when it failed. He announced that he wanted Henderson to head the eradication campaign. (The Russians, as the initial proponents of the idea, must have been angered by Candau's position.)

Now it was Henderson's turn to be unhappy. "The people I recruited for the West African program had been people that I had known from previous work, and many of them in CDC. I felt I really had a personal commitment to these people and had just gotten them signed up for this big program; some of them had left particular positions, and there was a certain personal loyalty to a number of these people. To suddenly pick up and go at this point was just unthinkable."

But Henderson was called to Washington to the office of James Watt, chief of the Office of International Health, a number-two man under the U.S. Surgeon General. Henderson remembers the conversation as going this way:

Watt ordered him to go to Geneva.

"In the Public Health Service," Henderson replied, "you just don't order people to go, you discuss these things."

"You're ordered!"

"Well, what if I refuse?"

"You resign."

"I don't like this."

"All right," Watt said. "We'll make a deal. Go over, do your best to get it started. I know it's difficult. I know there are problems. I know the prospects are not all that good. But get it started. You're assigned for two years, but if at any time you feel it is impossible just cable me, 'Now!', and I'll fish you out."

Henderson quickly discovered that one of his problems would be the entrenched bureaucrats at WHO. They were not geared to field work. It was largely a case of inertia. They also did not much like Henderson, who does not take obstruction quietly. Some did not like having an American running the program.

The budget for travel, for instance, was earmarked at $2,000 a year for Henderson and his staff of four medical officers and two administrators; this was patently absurd. It was not nearly enough to run a field operation. Henderson flew to Geneva (not included in the $2,000) and told the bureaucrats: "Forget it unless I can have $14,000." The bureaucracy was adamant.

Somewhat to his surprise, Henderson discovered some unusual allies within the bureaucracy. His principal support in WHO came from Karel Raska, a Czech whose wife was a member of the central committee of the Communist party of Czechoslovakia. The idea of the Czech director of communicable diseases fighting *for* an American apparently impressed many. In fact, Raska and Henderson had done some work together in the past.

"Raska just dug in his heels, and he is a very stubborn guy," Henderson says. "He wanted the job done. He felt I could do it better than anybody else, so he really fought."

Henderson got his $14,000 but had to agree to come to Geneva in November. The Hendersons and their three children moved to the city with something less than optimism.

A dispute soon erupted over research funds. The contention of the bureaucrats was that with the existence of an effective vaccine and knowledge of inoculation techniques no research was needed. Henderson fought and finally received "the magnificent sum of $40,000.

"But in fact, you could multiply the amount many times over, because if we gave, say, $5,000 to St. Mary's Hospital in London for research, at least ten times as much would be added by the investigators and the government. So $40,000 really purchased vastly more. But it was only after a considerable battle that I got any money at all."

The research budget item involved additional unpleasantness. Henderson was only thirty-eight at the time, ten years younger than any other WHO units chief with research-grant authority. His relative youth aroused resentment.

One of Henderson's first efforts was to develop a surveillance report, a publication that would keep track of smallpox outbreaks around the world. He had the initial issue published and was ready with the next a month later. Suddenly the smallpox unit found publication blocked. A high official did not want the report distributed.

Henderson wanted to know why. He was told that the publications committee had decided that the volume of material coming out of Geneva was excessive. Henderson explained to the committee chairman that the surveillance report was vital. The chairman said that he had not understood its importance and authorized Henderson to go ahead, but to change the name, because the committee simply couldn't reverse itself. The copy was resubmitted under a new name. Again publication was blocked.

"What's the trouble this time?" Henderson asked. "Well," he recalls being told, "there's a nursing report and a veterinary report and some other kind of report, and there's too much paper coming out of WHO. Why don't you use the 'Weekly Epidemiologic Record?' This was a publication of WHO which Henderson describes as a "laundry list" of all the afflictions of mankind area by area.

"It was out of date," Henderson says. "No one paid any attention to it anyway. If India was infected with smallpox you couldn't care if that little district in the middle of the state of Madhya Pradesh reported the disease or not. *India* was infected and you treated India as an infected country."

He bargained with the publications committee. Once again, he suggested someone else might do a better job.

It was eventually agreed that the smallpox unit could use the "Weekly Epidemiologic Record."

First he put the pertinent smallpox information on the back page of the record, "and as time went on we gradually pushed the laundry list of local infected areas to the back and put the smallpox up front. And then we stopped publishing the weekly laundry list every week; it became once a month . . . and they could only amend it during the interim weeks." Gradually the record became the smallpox report he originally wanted. It took almost a year.

He also issued a fortnightly publication to keep the staff informed of latest research findings. Every issue contained a scientific paper—"never mind what it was, or how good it was"—and the news was always upbeat. That, Henderson felt, was important.

Meanwhile the program was forming around the world. Since CDC was already at work in the twenty African nations, WHO would furnish it with support and it would continue under WHO auspices. Don Millar was running the program from Atlanta, with George Lythcott the head in Africa. Both had been chosen by Henderson and the relationship was easy.

Other African countries were staffed by experienced health officers. The French, British and Belgian colonial governments had had their own health programs—in the case of the French, very good ones. When Henderson was beginning his work France had already eradicated smallpox in a number of its former colonies. WHO could turn for staff to experts like Pierre Ziegler.

Ziegler had graduated from medical school soon after World War II and had joined the French colonial office in 1948 in Ubangi-Shari, now named the Central African Empire.* Three years later he went to Chad, where he worked for sixteen years. He never had a standard practice. His brand of medicine is a world unto itself.

Chad sits in the middle of the thickest part of Africa, landlocked, desolate, wretched. Once a part of French Equa-

* It used to be the Central African Republic until the President decided he liked being an Emperor better.

torial Africa, it became independent in 1960 but came to rely on France more after independence than when it was a French colony. Besides having almost no natural resources, it is constantly torn by the bitter conflict between the Arabs of the arid north, and the black Bantu southerners. French troops fought there for years, trying to establish peace and to keep Libyan-backed rebels from overthrowing the Bantu-dominated government.

In Chad—a country with a population of more than four million—Ziegler had six French doctors with him as the entire public health service, aside from nurses. They were in charge of the prevention of all communicable diseases, of which smallpox was only a small hazard. There were also sleeping sickness, venereal disease, tuberculosis and leprosy.

French doctors had been vaccinating in Chad for thirty years and generally had the disease under control. Most of the vaccination was performed by nomadic nurses, almost always male in the Muslim north, who were recruited from their bands, trained and returned to their people. They were given just enough vaccine and other medicines to last until the bands neared another town where the government kept stockpiles. The use of nomadic vaccinators would recur in Africa.

Most of the outbreaks of smallpox were imported from Nigeria, which borders Chad on the southwest. Nomads of the two countries constantly were crossing the border.

The nomads were not exactly sure what vaccination was for, Ziegler says, "but they were not against it. They accepted it. In fact, they understood this kind of product gave them some kind of protection against some diseases, but of course, it was not possible to explain exactly what smallpox was."

Ziegler kept track of his nomadic nurses as best he could. Frequently they would disappear with their people across the border into Sudan or Libya, only to return, their stock of vaccine and medicine long gone. Nevertheless, the last case of smallpox in Chad was in 1967. Despite the hardships, the disease could be conquered.

Ziegler was hired by WHO for its African program and was sent to Zaïre. That was one of the most difficult of all the countries to work in because of the geography and transportation problems.

"The problem of roads was one thing," Ziegler said. "The problem of bridges was worse. It was dangerous to cross some rivers, and very difficult on very poor bridges. The bridges were built by the Belgians before independence [in 1960] but maintenance was very poor after independence, and many bridges were blown up during the civil wars and then very poorly repaired."

"We managed to vaccinate a lot of rebels from the civil wars," Ziegler says, "a lot of old rebels. Also many from southern Sudan [which borders on Zaïre]. Rebels from south Sudan crossed the border and lived in Zaïre. There were no problems."

The government supported the WHO efforts and contributed what money it could. Ziegler had a Zaïrian counterpart who was "absolutely excellent. I think that before independence they had a very good health system, a very good health organization," Ziegler remembers. "There was a system of mobile teams going everywhere doing . . . sleeping sickness, yaws, tuberculosis, malaria."

But at independence the doctors, all of whom were Belgian, fled the country. The government of the country (then called the Republic of the Congo) asked WHO for help. The two hundred staff members sent out constituted the country's health system until 1967. WHO trained all the young native doctors.

All Ziegler's Zaïrian vaccinators were "locals," chosen by him or his staff and trained and supervised by them.

There was some resistance to vaccination in Zaïre, particularly among Jehovah's Witnesses. An arrangement was worked out whereby the Witnesses let the vaccinators "force" them into submission, which satisfied everyone.

Ivan Ladnyi, sent to head the program in eastern Africa, said that it was difficult to find "a common language" with some of the people, "I mean, difficult to convince them what is useful for them, what would be advisable for them," but eventually his efforts succeeded.

Another African country that gave vaccinators trouble was Uganda, where religious groups protested to the government.

The vaccine initially came from the United States. Soon huge shipments arrived from the Soviet Union, which would eventually supply the major share of the vaccine.

It was hoped that countries in endemic areas would eventually be able to make their own vaccine, but when two laboratories, one in Canada, the other in the Netherlands, tested some of the local products they found that barely 10 percent fulfilled WHO requirements. Some had no detectable virus at all.*

The 250 million doses needed annually for the mass-vaccination program would have exhausted the entire WHO budget. Accordingly all vaccine was donated. In effect this took all the vaccine available and ended any chance of a black market. WHO could put its money into the vital function of administering the vaccine.

Eventually local laboratories, under guidance from WHO consultants, began to produce usable vaccine, but for the first several years the program was based on the doses sent

* Some batches of Russian vaccine also were not very good. Henderson arranged a tour for several producers including the Russians at the Wyeth labs in Pennsylvania and to Moscow so that they could learn how to improve their vaccine. Even Wyeth had troubles getting the right potency at first.

by the Soviets and the United States, and a few other countries.

Relations between the field and D. A. Henderson's staff in Geneva were better than those between Henderson and some of WHO's Geneva bureaucrats. Ladnyi, for instance, based in Nairobi, Kenya, says that he could be in direct contact with Henderson. If there was any misunderstanding it was handled through channels, he says, something he considers very important to the success of the program. Eventually, however, it was Henderson at the top who resolved disputes.

"You cannot avoid political complications," Ladnyi says, "but at the same time using WHO headquarters as a buffer made it possible to unite our efforts." Ladnyi and others say that it was force of personality that kept Henderson one step ahead of difficulties.

"As a personality, he is very strong," Ladnyi says. "He does like people very much, but the proper people in the proper place and at the proper time. It was his technical knowledge, personal capacity and executive capability that produced such success. If it would be necessary to start again today, I think, personally, that I would recommend him." That is high praise indeed from the ranking Soviet official at WHO!

This understanding and cooperation between Americans and Soviets continued through most of the program. The smallpox campaign was one of the rare cases when the two countries proved that they could work with each other toward a common goal. Everyone tried to be considerate. The Americans never forgot that the Soviets conceived the program and were always careful to give them their due.

One American diplomat (he prefers to be nameless) who worked with both parties during the early years, said, "The very fact that the Russians had an interest in this early on made them feel entirely comfortable. This was not something that they were being dragged into, or had to defend at home. But if they didn't actually say that, they didn't need to, apparently. We would give them credit from time to time, and I think it was right that we did."

WHO continued to sign up countries. Brazil, director-general Candau's country, was more enthusiastic than he and quickly began a program. Countries of West Africa which were not original participants in the CDC effort also joined in. Indonesia started a program that would later prove crucial to eradication elsewhere when its lessons were absorbed.

But there were still some countries that did not consider smallpox a high-priority problem and held back. Many in WHO warned Henderson against even trying in a few countries.* Sometimes WHO's regional directors obstructed further participation.

Some of the obstruction involved sticking doggedly to the regulations. For instance, it was a WHO rule that no WHO activity would take place in a country unless that country spontaneously asked for it. Henderson found he had to send assistants around with a standard draft plan of operations and even would draw up the letter of request which the health ministers could copy to make sure they did it correctly.

"It was purely spontaneous," Henderson says with a grin. "Very spontaneous."

When three countries in the Middle East refused to admit they had smallpox, the smallpox unit reminded them how embarrassing it would be if a disease the countries did not

* Not a minor part of the warnings stemmed from the fact that communication and transportation in Africa at the time was very difficult. Telephone calls between the capitals of Dahomey and Nigeria, sixty miles from each other, had to be routed through London and Paris. There was virtually no east-west airline traffic either.

have spread beyond their borders. The countries finally admitted they really had smallpox after all. Regulations were violated,* but the fact remained that unless all countries participated in the eradication program, smallpox could not be eradicated. WHO could not leave a little pocket of the disease perking away somewhere.

Meanwhile WHO was getting considerable help from the big powers. Ben Blood was sent to Geneva and acted as a coordinator to make sure that the U.S. would provide anything needed. His job even entailed analyzing budgets of international organizations to make sure smallpox eradication was being adequately funded. The U.S. was prepared to give more money if necessary.

The Soviets had their representative present also. While they were less free with money, as noted they did make enormous shipments of vaccine.

The program was now in full swing. But it was becoming obvious that despite the victories, something was missing, something was not working.

* A WHO official in Geneva describes the policy this way: "We're an intergovernmental organization. WHO doesn't do anything in any country that any government doesn't like, good or bad. That's absolutely clear. We've seen examples of that with cholera. If a country says it doesn't have cholera, it doesn't have cholera."

XII

E^2 AND THE NEEDLE

While all the bureaucratic wars were being fought in Geneva, tens of thousands were vaccinating millions of people around the world. In many areas the disease was being controlled and even eradicated. Mass vaccination, with the jet injector or the conventional needle, was working in some places.

But in most endemic areas—the places where smallpox was a major health hazard—the program was not working at all. Smallpox was not being controlled in the Indian subcontinent, most of Africa or in Brazil, and no matter how hard the field workers applied themselves, they were barely holding their own.

It was estimated that between January of 1967 and December of 1969 a hundred million people were vaccinated in the twenty-country area of Africa patrolled by CDC-AID. In thirteen of those countries assessment teams discovered that

80 percent of the population had been vaccinated. But smallpox was still present.

The failure seemed to be with mass vaccination. As a policy, it was firmly implanted in everyone's mind as the right method. In 1964 WHO had declared that "the target must be to cover 100 percent of the population." Failing that, the eradicators believed that if they could vaccinate 80 percent they would get the upper hand. But that achievement did not suffice.

The programs also were clearly inefficient. The Indian government reported more vaccinations than there were people in New Delhi, but the disease continued unabated. Clearly, a lot of people had been vaccinated more than once, or a lot of vaccinators were lying. No one knew what to do.

Although no one realized it at the time, the first break was the epidemic in Nigeria that Hector Ottermuller had reported to William Foege. When the missionary radio network reported back to Foege the afternoon after his arrival at the epidemic site, they told him that they had found six more cases of smallpox. Four more were reported the following week. Foege, short on vaccine and on personnel, responded to each report by sending all the people and vaccine he could to the site of the outbreak. He gave up any notion of trying to vaccinate everyone in the province.

The second week after the initial outbreak there were twelve new cases, and in the week after, nine. Then the disease disappeared. The vaccinators found that, except for two people they had missed on their first sweep, the chains of infection had been broken in three weeks.

Only a small minority of the people in the affected area had been vaccinated, yet the disease had been beaten. It took a while for that lesson to sink in.

Meanwhile, further epidemiologic studies turned up some surprises about smallpox.

"The textbooks kept saying that this was one of the most contagious diseases," Foege says. "We believed them until we started the epidemiology. We went to villages where smallpox had broken out and found that there hadn't been a case for fifteen or twenty years. This is unlike measles, where every other year you expect an outbreak. So gradually we developed the idea that smallpox really comprises islands of infection that move around slowly, but is not something that covers an extensive area like a country all the time. During the smallpox season in West Africa maybe only 1 percent of the villages contain cases at any one time."

Studies in India showed the same ratio.

Rafe Henderson says, "We had instances of rural populations where we have documented transmission continuing over six or seven months with one little nomadic group traveling across the desert. You'd think that one group or one member of the group contracted the disease and that in the next generation of the disease virus all the rest of the susceptibles came down with it. But that wasn't what happened."

One person would get the disease, pass it on to one or two others, who passed it on similarly in turn. The chain would continue for months, and thus survive in sparsely populated areas. Instead of blowing itself out in a small group by infecting everyone all at once, it hung on for months, guaranteeing that sooner or later it would encounter another group of susceptibles and insure its survival for several months longer. This characteristic made the virus vulnerable to attack, however, because the chains of transmission were tenuous and complex. They could be interrupted and contacts could be traced.

"With the slow spread you have time to do something," Foege says. "It might take three, four or five generations of the disease to get through one household, much less a village. So, if you set up a surveillance system to let you know where smallpox is, you can actually get in there and do something."

Another conclusion was reached by the CDC people: there was a season for smallpox.

In 1966 Foege had drawn up a series of maps that showed where and when outbreaks had occurred in eastern Nigeria. He found that each year, usually in the late summer and autumn, there was a period of very low incidence. The epidemic season was the spring. The epidemics usually began with a few outbreaks in the north and month by month they would creep south. By late spring there would be an epidemic over all the area.

The CDC-AID people were responding with everything they could find. As reports of an epidemic came in, they rushed all their resources to the area. But they never could keep up with the virus.

Foege and several of his coworkers realized that they were not making the best use of their resources. Since there was a period when the disease was relatively uncommon and most vulnerable to attack, that was the time to try to contain it. Then it might be possible to break the individual chains of infection, which could not be accomplished during the epidemic season.

Studies in Pakistan also showed fluctuations usually riding on the annual monsoons. Spring, preceding the heavy rains, was smallpox season. Autumn was not.

One more lesson was to be learned: When an outbreak did occur, *everyone* in the area had to be vaccinated. An unprotected pocket would be found by the virus.

This became abundantly clear in the Nigerian town of Abakaliki, where CDC-AID launched a massive vaccination program as a test of its ability to cover urban areas. After a concentrated assessment, they found a vaccination coverage in excess of 90 percent. However, at the Faith Tabernacle, a religious community, 120 people refused vaccination. Thirty of them came down with smallpox. Coverage in infected areas, then, had to be total.

Foege, Don Millar and CDC's Mike Lane decided to test their belief that the time to break the chains of smallpox transmission was during the off-season. The government, which had been exceptionally cooperative, agreed to assist an investigation extending across eastern Nigeria. The de-emphasis on mass vaccination and striking during the autumn made smallpox disappear in five months. CDC-AID had vaccinated only 750,000 people out of a population of over twelve million, a little more than 6 percent.

The results were not immediately apparent because, as the last cases were reported, civil war broke out. The best clue of success, however, was that, although part of the region covered by the vaccinators was in rebellious Biafra, no smallpox was reported in any of the war areas.

The war forced an evacuation of the missionaries, and Foege's mission was quickly overrun. He was at a meeting in Accra, Ghana, at the time.

When the success became evident, the epidemiologists decided to expand the experiment. The program now was called "Eradication Escalation," or "E²." Eight countries were chosen: Nigeria, Togo, Dahomey, Mali, Upper Volta, Niger, Guinea and Sierra Leone. They began in September, 1968. The whole tactic centered around good surveillance—finding the outbreaks—and containment—checking their spread by total vaccination in the outbreak area. *Surveillance* and *containment* were the watchwords.

Instead of the usual increase in winter cases building up to a large epidemic in the spring, cases kept declining as the reporting system improved. Within a year smallpox was eradicated in most of the eight countries. In Sierra Leone, which had begun its program only in 1968 and at the time had the world's highest rate, Don Hopkins and his crew wiped out the disease within a few months. They used surveillance and containment only—never mass vaccination—so the victory was unsullied. Only 66 percent of the population of Sierra Leone was ever vaccinated in the program. In Mali the disease was eradicated with only 51 percent vaccinated.

The key in many ways was the surveillance. In the first six months of 1968 WHO estimated that less than 5 percent of all the cases were being detected and reported. By fall 57-67 percent were found in West Africa, and by January of 1969 every single outbreak in the eight-country area of the E^2 experiment was reported and responded to.

Once the surveillance-and-containment principle had been adopted the eradicators realized that it had been staring them in the face for years. They were not the first to try, only the first to put a name to the practice and to apply it on a large scale.

It was found, for instance, that in 1896 a commission in Great Britain trying to find out why there was still smallpox in the country despite widespread vaccination called for "a complete system of notification" to pinpoint outbreaks so that there could be instant reaction. In his classic 1962 book on smallpox, C. W. Dixon wrote: "If more study was given to the foci of smallpox, it might be possible to eradicate the disease from an area by vaccinating a far smaller proportion of the population."

More to the point currently, the principle was being proven simultaneously in Indonesia, where a WHO-supported program had begun in December, 1967. There too mass vaccination wasn't working.

That year a meetıng was held in Bangkok for all the countries in the southeast Asia region. "The regional director for WHO was bitterly against the program," D. A. Henderson says. "His attitude was: Nothing else works in this region and this won't either, forget it."

An Indonesian, Ignacio Setiady, Henderson and Jacobus Keja, a Dutchman, met in the dining room of a Bangkok hotel trying to see if they could get the program on its feet in that nation of many islands. The Indonesian outlined all the problems, not the least of which was that the vaccinators were getting only $1 a month in pay. Henderson, hoping to free more Indonesian money, pledged $100,000 from WHO to help pay for the program.

"They began their expanded program in July of '68, and the great, tall Dutchman, 'Ko' Keja, was out there frequently," Henderson says. "They did surveys to get some idea what the immunity level was and found in central Java that it was . . . 95 percent. And they had good vaccine—it was being produced locally."

Yet, still there was smallpox.

"So that was a puzzle. . . . We decided that with 95 percent vaccinated, we should be containing outbreaks. The Indonesians didn't agree. They were determined to have everybody vaccinated.

"So they set up a huge program in central Java, God knows how many 'backlog fighting teams,' as they called them. They would go village by village, find out all the people who weren't vaccinated and vaccinate them," Henderson remembers. "This was a crash effort. When it was completed they still had a lot of smallpox."

In desperation the Indonesian government set up four "firefighting teams" simply to go after the outbreaks in central Java. "Literally, within months, we went from several hundred cases a month to zero. We stopped it," Henderson recalls.

The surveillance-and-containment technique was expanded throughout Indonesia, one of the most difficult countries in the world to operate a public-health program.

"You have to walk up mountains and walk through rice fields to get to certain places," says WHO's Nicole Grasset. "You've got to go on lakes where there's terrific winds, and take risks. There are many islands, of course." Many of the vaccinators would go island hopping by boat even though some of them did not know how to swim.

The vaccinators were dispatched by the central smallpox unit in Djakarta which was run by the Indonesian version of CDC. They were simply told not to return home until they had finished their work. Since they were sent far from home, there was an added incentive to complete the work quickly.

WHO wanted to encourage cooperation by issuing a reward to anyone reporting a case of smallpox and by paying vaccinators more. The government had objections to both proposals. One reason that WHO wanted the higher pay was that many local health workers had to moonlight in order to survive and were distracted from their smallpox work.

"I remember, for example, a man . . . in Makasar, the capital of South Sulawesi on Celebes," Grasset says. "He was in charge of the municipal smallpox-eradication program. That man was not receiving anything from us. . . . But he was achieving smallpox eradication . . . he was also in charge of certain wards in the hospital, he had a fellowship from WHO for research, and he had a shop and a private medical practice. He had five activities, and he had to, because he was not earning much. All the health staff in southeast Asia were making very low salaries. . . .

"When we gave them a per diem and a salary it helped some to give up other jobs and they were able to work for the program full time."

The vaccinators in Indonesia ran into frequent resistance against vaccination and sometimes had to call in the police. The policy was to vaccinate everyone, even the newborn.

"I remember I went to a village after there had been a big epidemic," Grasset recalls. "Everyone had been vaccinated except one baby who had been born two days before. The mother didn't want vaccination. . . . There was no more smallpox in the village except for one child just twenty-five days old who was full of the disease." The younger infant caught smallpox from the older one two or three days later and died.

"Having seen this in the first month, I was convinced. When I later ran into mothers in India who didn't want their young children vaccinated, I'd show them a card with a picture of an infected child and say, 'Look at your beautiful baby. Do you want him to become like this? He could even die.' One woman said to me, 'It doesn't matter if he dies. I'll have another one next year.'

"What can you answer to that?"

By January of 1972 smallpox had been eradicated in Indonesia.

There was one more country that bolstered the case for surveillance and containment, WHO director-general Marcelino Candau's Brazil.

"We had a helluva time with the Brazilians," D. A. Henderson says. "They were basically malariologists, and they ran the smallpox program much like the malaria one: If you sprayed all the houses malaria will stop, so you spray all the people and smallpox stops."

In Montevideo, Uruguay, Henderson met with the Brazilian health secretary and persuaded him to try surveillance and containment. The secretary agreed to set aside three medical officers, each with a vehicle, a driver and a vaccinator. "That was to be the whole surveillance effort."

One team was placed in each of three states, Paraná, Minas Gerais and Bahia.

"The fellow in Paraná was really something," Henderson says with affection. "Ciro de Quadros. I remember visiting him with his driver and vaccinator. He got information from every damn health unit and hospital reporting cases of smallpox, and then the three of them would go out and do the containment vaccination.

"I think, if I recall correctly, that in the space of nine months they investigated and contained outbreaks amounting to over a thousand cases. At about that time the systematic vaccination program started and went east to west across Paranâ, and we found no cases.

"I saw one of his reports of an outbreak. Well, it was so beautifully written up you could reference it, put an introduction on it and publish it. It was beautiful. I told him, 'My main criticism is that it's very nice to write these things up but you can't take that much time writing and get the job done.'

" 'Well,' he answered, 'you know I have to come back once a week, and if I come in on Saturday night, I can begin writing then, get my supplies on Sunday, finish writing Sunday night and be back in the field Monday morning.' "

With people like De Quadros Brazil eradicated smallpox in 1971, the last country in the Americas to achieve this feat. Smallpox, which had crossed the ocean with the conquistadores, was driven back with the vaccinating needle and the surveillance-and-containment campaign.

In 1972 the technique was formulated in a scientific paper that Foege, Millar and Lane published in the *American Journal of Epidemiology*. Titled "Selective Epidemiologic Control in Smallpox Eradication," it was perhaps one of the most important documents in the campaign against the disease since Jenner. It laid out the case for a complete reversal in techniques. It concluded:

> "Selective epidemiological control" has been successful in West and Central Africa. Intelligent use of vaccination based on knowledge of where the disease is and when, where and to whom the disease is likely to spread, is more economical in time, vaccine, and personnel than blind mass vaccination. Mass vaccination campaigns should continue in endemic areas, but we

> consider the use of surveillance, investigation, and selective epidemiologic control techniques to be of equal and, under certain circumstances, of even greater importance than systematic mass vaccination activities.

The problem clearly was getting everyone else to agree with this most drastic change in a policy two centuries old.

Simultaneously, there was another revolution taking place, this one concerning the lowly vaccination needle.

Since Jenner the standard instrument of vaccination was the single-pointed needle. The conventional technique was to put a drop of vaccine on the skin and then, holding the needle parallel to the skin, penetrate about twenty times with the point. The depth was just enough for the vaccine to enter the skin. The excess vaccine rolled off the arm or was wiped away, wastefully. If done properly there would be a small red mark but no blood.

In many cases not enough vaccine got into the skin to produce a "take" and the failure rate was high. Many American doctors wanted the "take" to be as small as possible for cosmetic reasons. They did not understand that unless sufficient vaccine were introduced into the body, there would be no "take." Quite a number of people who had been "vaccinated" actually had very little or no immunity to smallpox. Either the needle or the physician failed.

The U.S. Army's jet injector was a great help. Take rates ran almost 100 percent and thousands could be vaccinated in a day if they were assembled. The problem with the jet injector was that it is a relatively complicated mechanical device, prone to malfunctions. In the bush in Togo, or a slum of New Delhi for that matter, spare parts are hard to come by and a broken injector is useless.

What was needed was a needle that would hold exactly the dose for one vaccination, be capable of use by an amateur and never miss.

Such a device came out of Reading, Pennsylvania.

In 1961 Dr. Benjamin Rubin, a researcher for Wyeth Laboratories, located near Lancaster, began to experiment with alternative devices to the conventional needle. He consulted with Gus Chakros, an engineer with the Reading Textile Machine Company, on the design.

Rubin was looking for a needle that would hold the proper amount of vaccine by capillary action, the way a tiny loop holds a bubble of water. He took a sewing-machine needle and ground down the eyelet end into a fork shape. Chakros said he could do better. He cut a piece of thick wire to the right length and then stamped it to create the fork, or bifurcation, that would hold exactly one milligram of water between the two sharpened prongs.

Chakros and Rubin tested the needle with the reconstituted Dryvax vaccine, a freeze-dried product developed by Wyeth's Dr. Howard Tint the year before, and found that it held perfectly a little more than one milligram of vaccine (the difference results from Dryvax being denser than water).

Now it was time to test the bifurcated needle on humans. That chore fell to Dr. Malcolm Bierly, who went to the juvenile detention center in Camp Hill, Pennsylvania, outside of Harrisburg. New inmates were required to be vaccinated against smallpox. The first test, performed on thirty-one youngsters with the prison officials' consent, was spectacular.

Each youth was vaccinated twice on the same arm, with the regular single-pointed needle and with the bifurcated needle. The "takes" with the bifurcated needle were twice the size of those with the other needle every time.

In his patent application Rubin wrote: "Dramatically, the

vastly superior results were obtained with the forked needles despite the fact that the conventional needles utilized 20 times the amount of vaccine required with the forked needles and [the conventional needles] were pressed against the skin four times the number employed with the forked needle."

If Bierly, a known skeptic, needed to be convinced any further, a silly accident in Wyeth's dispensary resolved his doubts. As he was vaccinating employees with the bifurcated needle, he tossed one needle toward a wastepaper basket. At that exact instant a nurse was throwing something into the basket and the flying needle grazed her hand. Three days later she came down with a vaccination reaction at the point grazed.

Bierly continued testing and ended with more than a thousand vaccinations with 100 percent take rates. Later he found that pressing the needle to the skin twice was enough to get 100 percent rates in children's primary vaccinations, when the child has no antibodies. Revaccinations required only four to eight presses to get a take. It needed twenty penetrations to get this result with a conventional needle.

The researchers could get takes even with vaccine diluted ten times. "This immediately produced a tenfold increase of the world's supply of smallpox vaccine," Bierly says, "a major factor in the eradication." *

"It was [even] demonstrated that this method of vaccination would give the maximum expected response if *any* visible amount of the standard vaccine was seen to adhere to the prongs of the fork. This system was therefore found to be quite foolproof."

A patent application was filed on October 5, 1962, by Rubin with an assignment to American Home Products Corporation of New York, the conglomerate that owns Wy-

* WHO officials say the actual savings were only about fourfold.

eth. The patent, number 3,194,237, was granted July 13, 1965.

Chakros devised a method of quantity manufacture, and all that remained necessary was for Wyeth to determine the packaging and marketing. For sale the company designed kits that contained both doses of Dryvax and bifurcated needles. This seemingly straightforward approach would lead to difficulties.

Shortly after taking his position in Geneva, D. A. Henderson wrote to Bierly praising the needle as a "truly imaginative invention" and suggesting that it would be as great a boon to his smallpox eradication program as the jet injector. He told Bierly that he was field-testing the needle to see if it was as efficient as Wyeth claimed.

Field test showed that the needles were even better. WHO purchased hundreds of thousands and shipped them to its vaccinators. Ziegler used them in Zaïre as early as 1968 in place of conventional needles. "The bifurcated needle is absolutely perfect," he reported.

But Wyeth was not maximizing its profits. It had granted WHO free rights to manufacture the needle, but the international sales department otherwise wanted to sell the combined needle-vaccine kits. If the buyer wanted the needles, he had to buy Wyeth vaccine. That insistence touched off a battle that lasted for years.

On June 18, 1970, for instance, J. H. Brown, managing director of Wyeth in Marietta, wrote to Henderson:

> It has been a long time since we discussed the availability of our bifurcated needles to European producers. This subject has been one of long discussions in our Executive Committee. While some of us were pushing to make the needle available to everybody, the group—principally the International Group—holding out for re-

> stricted use, have won the argument. We have had a number of inquiries from European manufacturers and on the basis of this decision have found it necessary to deny them the use of the needle. In making it available to the World Health Organization, it is felt we have at least in part fulfilled our moral obligation.

There are some critics of the industry in WHO who believe that discussing a "moral obligation" with a drug company is like trying to discuss birth control with a rabbit; there is at best a language problem. But in fact Wyeth had been quite liberal with WHO. It was the matter of sales to countries wanting to use the needles themselves that was at issue.

Henderson shot back a letter to Brown, pointing out that this little bit of economic blackmail would not work:

> I was deeply chagrined to receive your letter of 18 June and to learn that the needle would not be made widely available. There are, as you have suggested, a number of European health services which are interested in obtaining them. I am equally confident, however, that they will not buy vaccine to obtain the needles as virtually every country has its own national production center. Various vaccination instruments have been used for many years and everyone is reasonably satisfied (as they should be) with what they have used and thus there is no great pressure to change. I wonder if the Executive Committee is aware of the fact that there is today virtually a nil commercial international market for smallpox vaccine. With the eradication programme, we have assumed the obligation of providing vaccine, either donated to WHO or by bilateral channels, to the endemic regions. A number of

> European countries have constructed or expanded their facilities for freeze-dried vaccine to handle their own needs and for donation. . . .
>
> Perhaps there is a rationale I don't comprehend but, from this vantage point, it would seem to me that either money might be made by selling them as a separate item or they simply won't be sold at all. I'm no business man but I think I know the choice I would make.

Henderson knew better than Wyeth just how good the needles were. He added a paragraph to his letter recounting the success of the bifurcated needles, the ease with which people had been trained to use them and the fantastic take rates.

In February, 1971, Wyeth's sales department appeared to back down. Brown wrote to Henderson:

> You will be pleased to know that on Friday of last week, our top management reversed themselves on their position re the bifurcated needle. The decision has now been made that the needle will be available in Europe, Asia and places other than the United States. The efforts of Drs. Bierly and Tint and others in beating the drum did bear fruit.

Brown added that Wyeth would do everything possible to keep the cost of the needles down.

Unfortunately, the word did not get passed to everybody at the giant drug company. On May 11, 1972, Henderson wrote to Tint. He complained that a number of European countries had been turned down in efforts to buy the needles without purchasing Wyeth vaccine. He pointed out that they were manufacturing their own vaccine and would rather give up the needle than make unnecessary purchases.

WHO's policy of taking vaccine off the world market

meant that Wyeth's sales department was using a priceless piece of equipment to force the sale of a commodity that had no value.

Tint wrote back on May 23, expressing surprise that there was still a problem and assuring Henderson that "under current sales policy there can be the sale of needles without accompanying vaccine."

The matter was thus resolved. WHO, which had been granted the licensing rights free of charge, had a German company make the needles at a lower price than Wyeth's. The needles soon became the symbols of the smallpox eradicator. Nomads were taught to use them in twenty minutes. Indian vaccinators carried dozens in canvas bags through the parched plains of Uttar Pradesh. And, in fact, Wyeth made very little money out of the invention, another example of how the eradication program somehow managed to bring out the very best in people.

"I've spent my life in this business," says Rubin, "and I've done a lot of things that took a lot more effort and thought, but I guess this proved to be the most effective."

WHO would refine the technique with the bifurcated needle, making it even more efficient. It was found that the best way to use it was to hold the needle at a 90° angle by having the vaccinator rest his wrist on the patient's arm. The vaccinator then jabs downward about fifteen times in an area about fifteen millimeters in diameter. There should be just a drop of blood here and there, nothing profuse, but the blood is a sign that the needle is getting deep enough. Bierly insists that there is no data to prove that this method is best, but it is easier to explain to nonprofessional vaccinators.

Sterilizing the vaccination site was shown to be a waste of time. When the program began it was common to wipe the skin clean with a 70 percent alcohol solution. Then dispensing with sterilization was tested. Caked dirt was wiped off, but otherwise nothing was done to clean the skin. The secondary-infection rate was no higher than before. Henderson figured out that "the savings in alcohol, soap and cotton sponges paid a substantial proportion of the gasoline costs in some places."

To handle the needles WHO designed a plastic container that, when shaken, permitted the vaccinator to take out one sterile bifurcated needle at a time. When the needle was used it was placed in a similar container (usually of a different color) and inside a metal box filled with water and boiled. After a half hour the needles were ready to be used again.

The vaccine itself was kept dried in a small glass vial. When the vaccinator was ready he would take a small glass cutter and open the nipple at one end. From a similar vial some glycerol was dropped into the vaccine. This combination provided one hundred vaccinations with the bifurcated needle in almost completely sterile conditions.

The bifurcated needle, a vaccine that was 99 percent effective and stable, and the surveillance-and-containment technique were the weapons needed to conquer smallpox.

In June, 1970, there was to be a dinner party in the Nigerian town of Kaduna to honor the Nigerian vaccinators and workers who were being released from the concentrated phase of the program to participate in the maintenance phase. As far as anyone knew smallpox had been eradicated in West Africa. Now the workers were to watch for importations of the disease from elsewhere, to vaccinate and to conduct the measles work as AID had originally intended.

Just before the party was to begin a man rushed in from the infectious disease hospital—a hut on the outskirts of town—claiming to have seen a smallpox case.

"Everyone's mood went down the drain," Stan Foster says.

Several people raced to the hospital. On the front porch was a case of chickenpox. Everyone laughed.

"That's not the case I asked you to see," the man rejoined.

Around the back they found an adolescent girl obviously suffering from smallpox. She had come to Kaduna on the train just a few days before.

The vaccinators went to her village and asked the headman if there was any smallpox in his village, any deaths. He said there had been none. They kept asking, and he kept denying that his village was infected.

The conversation was interrupted when the headman's granddaughter walked out of a hut. She was covered with smallpox.

Sixty to seventy cases were found in the area and a house-to-house vaccination program began.

That was the last outbreak of smallpox in West Africa. The campaign that began with Larry Altman and his exploding trucks had ended in complete victory.

The scene now shifted to India, Pakistan and Bangladesh. Here smallpox may have been born, here it was indelibly intertwined with the daily life of the teeming subcontinent, here it would be the hardest to kill. And here there was a goddess to slay.

XIII

Encounter with a Goddess

Mother has come; she has come from the forest,
She throws pearls from the cart, Oh Mother Mata.
Mother has come from another region.
She throws pearls from the sugar cane garden,
Oh Mother Mata . . .
Mother travels at midnight,
In every house Mata's cots are swinging,
Oh Mother Mata.

She is naked and sits atop a donkey. She carries a broom or a whisk in one hand, a pitcher in the other. On her head is a flat basket, the kind used to sort grain.

She is Sitala, or Shitala, or Mata-May, or Gangamma, and she is the goddess of smallpox.* She is short-tempered

* She is actually the goddess of smallpox and allied diseases; she is responsible also for measles and other pox diseases.

and cruel. A believer never says anything negative about her. One does not say, for instance, that she causes smallpox, although to believers it is clear that she does. It is said, rather, that she is merciful because she spares many of her victims. The sickness is a sign of her anger, recovery of her mercy.

She is naked because usually her victims cannot wear clothing, the sores are so painful, the heat so unbearable. The uses of the broom or the whisk differ, depending on which legend is being related. Some say that the broom is to sweep up the victims, others that the whisk is a whip that raises the pox marks where it hits the skin. The pitcher contains cooling water, the balm of those fevered by the disease; Sitala means "cool one."

The basket contains either pearls or seed. The goddess strews them about, and where they land they cause a pox mark. Others maintain that the basket holds the victims swept up by the broom.

Like most things in India, the legend varies.

Many authorities believe that smallpox evolved into its final form in India or China and spread around the world from there. If so, India is the world's oldest endemic country and the disease has been so much a part of everyday life that it has been enshrined in religion and folklore.

Oddly, very little is known about the smallpox in India before 1900, because the British did not keep good records until then and the scores of India's conquerors who preceded them kept none at all. The disease seemed to run in five- to seven-year cycles. It is known that the strain was the pure, terrible variola major. The death rate was probably near 20-30 percent, high enough to encourage anyone to pray for relief.

Even after the British began keeping records, underreporting—a problem that would haunt the eradication effort—was marked. Current estimates are that in areas where the health services were relatively disorganized only about one case in every hundred was reported and in better organized areas the ratio may have been one in twenty-five to fifty.

The overall annual death rate was probably the highest in the world. Between 1900 and 1920 it averaged 370 deaths for every 100,000 people, with a high of 800 per 100,000. The rate improved after 1920, but was still among the world's highest. The incidence of the disease was greater than in any other endemic part of the world when WHO began its expanded program in 1966.

Leaving aside precise figures, what we do know is that as recently as five years ago there were epidemics so terrible that rivers were clogged with dead bodies.

Even optimists at WHO and in the Indian government—people who firmly believed that smallpox eventually would be eradicated—were convinced that India would be the last country in the world to achieve the goal.

"India is not Nepal, India is not Sri Lanka [Ceylon], India is not Pakistan, India is not Afghanistan, India is India!" says Dr. M. I. D. Sharma, one of those who would head the effort.

Indeed it is.

It is a country of more than 600,000,000 people, 1,652 recognized languages, three distinct racial groups, six major religions and several thousand variations. It runs from the foothills of the Himalayas, through the Ganges plain and the Deccan plateau, passes the scruffy vastness of the center, to the ocean and lush forests of the south. There are 580,000 villages and 2,648 towns and cities, 147 of which have more than 100,000 people and nine of which have more than one million.

The total population is second only to China's. Population density is among the highest in the world. The poverty of most of the population is too well known to require emphasis.

India may be the most complex nation in the world—not really one country, but hundreds, all gathered under one flag. It may also be the most complex society on earth. No one really understands India as an entity, not even the Indians. As soon as one truth becomes evident, it is supplanted by its opposite. India, as Sharma says, is India.

To believe that a smallpox-eradication program could be organized that would visit every village, every hamlet, every person and record the results was insanity. But it was done.

The goddess Sitala placed some obstacles in the path of the would-be eradicators. Many of her worshipers believed that vaccination would incite Sitala to wrath against those trying to resist her will. Vaccinators had to try to convince these devotees that the goddess would not disapprove.

But in some ways the worship helped slightly to limit epidemics. Certain rituals effectively isolated victims from others who were highly susceptible. Because smallpox was essentially a childhood disease in India, most of Sitala's devotees were women, and they performed the rituals devotedly on behalf of their offspring. The mother would place the child victim in a separate room. If the house had only one room, he would be along a wall, either on a cot or on the ground. A sheet would be hung from the ceiling to block off the patient.

To warn others of smallpox, the leaves of the *neem* tree were placed over the doorway, a universal sign that Sitala was within. Generally no one not belonging to the family, particularly children, was allowed to enter the house. Outside the village a red banner was hung across some poles to keep strangers out, including doctors and vaccinators.

When one entered a smallpox house shoes had to be

removed. Since the goddess was present, the house was a temple. Hands were washed on entering and leaving.

The victim was not bathed until the scabs fell off, and his clothing and linen was not sent out to be washed—a useful practice in avoiding contamination.

The fevered victim was fanned with *neem* leaves. In the hottest parts of India they were placed on swinging cots where they could be cooled by breezes. There were many references to cots in the Sitala liturgy. When the poet wrote, "In every house Mata's cots are swinging," he was alluding to a smallpox epidemic.

In 1966 a British sociologist published the results of a survey made in a northern Indian village. She asked the villagers what they thought caused smallpox. More than 55 percent said the goddess Mata. Only 19 percent blamed "unseen living organisms." * Smallpox vaccinators would have the goddess to contend with.

When Zhdanov's resolution was passed by the World Health Assembly in 1958, a number of countries, including India, were inspired to begin their own programs. But getting anything started and working in India takes time and considerable effort. The Indian government and its bureaucracy are astonishingly cumbersome even for so large and complex a country. Traditionally in India there are twenty people hired for every single job. Employment takes precedence over efficiency.

In part the federal form of government is to blame for the bureaucratic difficulties. Although the central government works on the British parliamentary system, the country has twenty-two states and nine union territories. Under the Indian constitution a number of powers are reserved for the

* Four percent blamed evil spirits, 2 percent foul smells and 17 percent said the cause was unknown. The rest had no opinion.

states that elsewhere might be within the purview of the federal government. In that respect it resembles Canada more than either Britain or the United States.

One of those reserved powers is the regulation of health.

In the federal government there is a minister of health and family planning. His principal responsibilities are general health, acting as liaison within the government and with international organizations (such as WHO) and establishing federal policy regarding health research and education. Under him is the secretary of health, his executive arm, and the directorate general of health services, which supplies technical assistance.

But it is on the state level that the health services are run, and here the complexity and enormousness of India come into play. Most of India's states could make large countries all by themselves. Uttar Pradesh in the north had 88,341,144 people in the 1971 census, which would make it the eighth most populous country in the world if it were independent. Two states, Bihar in the east and Maharashtra in the west, have more than 50 million people. Some of the states are well run, some not. Some couldn't be well run under even the best of circumstances. They are too big, or too crowded, or too diverse in population, or too poor, or too independent, or some or all of the preceding. But it was with these state governments that the eradicators had to work.

The state governments were set up along the same lines as the central government. Each had a minister of health and the usual subordinate executive branches. States were divided into districts, each with a chief medical officer and at least one hospital. Below the district level was the primary health center (PHC). Each PHC has at least one physician and is responsible for 80,000-150,000 people and 150-350 villages. The PHCs are the basic health provider for the people. It is in them that government and the people meet and there that government would have to be at its most efficient if smallpox was to be eradicated.

In May, 1958, the central government minister of health and family planning appointed a central expert committee to examine the smallpox problem and to make recommendations for an eradication program. A year later the report was issued. Its recommendation: the establishment of a National Smallpox Eradication Program (NSEP) to vaccinate *everyone* in India in three years.

The committee suggested pilot programs to find the best method of operation. Such programs were set up in October and November, 1960, in one district of each state and in the union territory of Delhi. According to government figures, there were twenty-three million people living within those pilot districts and more than twelve million of them were believed vaccinated by March, 1961. Those figures—like most figures in India—were probably fanciful.

In October, 1961, it was decided to expand the program throughout the country at large during the Third Five Year Plan.

A total of 68,900,000 rupees * was set aside for NSEP. The Soviet Union agreed to supply the vaccine, beginning with 250 million freeze-dried doses. The Soviets would double that by about the end of 1967. In October, 1962, NSEP was launched across India.

Five months later WHO and the U.S. Agency for International Development (AID) were invited to assess the program. Another assessment by India's National Institute of Communicable Diseases (NICD, its version of CDC) began shortly thereafter. Both assessments were critical. NICD, for instance, found that in six districts more vaccinations were reported given than there were people.

"The Indians, in a classic problem of a large bureaucracy, had hired a number of very low-paid vaccinators," says one of the Americans involved in the assessment. "They had a

---

* A rupee was worth 12¢ then; it is worth about 8¢ now.

quota. So, of course, they vaccinated and revaccinated the very easily accessible population. And there were some 10 to 20 percent of the Indian population that was at high risk of smallpox, that were difficult to get to, who'd run away if you came at them because they didn't want [vaccination] sores, who were unvaccinated. Smallpox just continued in that unvaccinated segment of the population. It was a good example of how an ill-run mass-vaccination campaign could be quite cost-ineffective."

They were other errors as well. In some areas the vaccinators were part of the same bureaucracy as the tax collectors, which did not encourage cooperation. Sometimes a member of an unacceptable caste was sent into a village.

Nonetheless, in January, 1964, a central council on health ordered the vaccination of 100 percent of the population.

By the time in 1966 that the World Health Assembly had the funds to proceed with its eradication program, India claimed to have performed 60 million primary vaccinations and 440 million revaccinations. The same year it reported 80,000 cases of smallpox, the same number reported the year the Indian program began. And that figure represented less than one case in 20. In October, 1967, the Indian government called for help, and a joint government-WHO team assessed the program and admitted that, so far, it was a failure.

The Indians also were being pressured by the Soviet Union. For their half billion doses of vaccine the Soviets demanded reforms. India asked WHO for additional help. WHO sent four medical officers.

The army began to form.

When the first four officers arrived they were put in the charge of Nicole Grasset.

The daughter of a Geneva microbiologist, Grasset holds dual citizenship, Swiss and French. She was trained as a dermatologist and venereologist and studied tropical medicine in London. Early in life she decided to dedicate herself to helping the helpless of the world, a goal to which she brought boundless energy and impatience. When a London appointment to work in Ghana fell through, she left England and joined France's famed Pasteur Institute in Paris "for a couple of months." She stayed there ten years, working mostly on rabies and measles but never forgetting her commitment.

When the Nigerian civil war broke out she could not stand by. She contacted the International Red Cross and asked to be sent to Biafra to help those caught in the war. WHO declined because it could not recognize Biafra, a rebellious portion of a member nation. She turned to D. A. Henderson, for technical help, then taking up his position in Geneva. Neither could he help her officially, but he did give her advice and assistance in getting bifurcated needles and vaccine. Grasset wound up flying to Nigeria from Geneva on weekends for the International Red Cross, and spent several weeks bringing medical supplies and vaccinating for several diseases in the war-torn Nigerian villages.

Henderson knew immediately that he had to have Grasset on his smallpox team. He asked her to go to Indonesia and then to assume the top operating position for Asia in New Delhi.

Grasset, then around forty, looked at least ten years younger. She is a beautiful woman, always perfectly, tastefully dressed. Like many European middle-class women, she had only about five suits but each was kept in perfect condition. Although she went infrequently into the field in India, she would be dressed then as if for the office. Several cherish memories of Grasset touring a village in India, making no concessions at all to the heat or filth, looking cool and dignified.

When she arrived in New Delhi she was one of two smallpox medical officers in WHO's slightly seedy five-story building located across the street from a belching electric-power generating station. The other officer eventually was withdrawn and for more than a year she was alone. Her partner on the Indian side was Dr. Mahendra Singh of the health ministry. The two were the only full-time smallpox officers when WHO's first four epidemiologists arrived.

Each one was assigned with sole responsibility to a state or a region. "They had a terrible job," Grasset says.

A Ukrainian, V. A. Moukhopad, was assigned the entire state of Uttar Pradesh with its 100 million people. Albert Monnier, a Mexican, was assigned to Rajasthan, the forlorn desert state in the northwest bordering Kashmir and Pakistan. An American, John Pifer, was given crowded Bihar in eastern India. A Czech, Vladimir Zikmund, was sent to southern India, where the disease was not endemic but where cases constantly were being imported. The group of four later was joined by another Russian, Slava Selivanov, who was sent to West Bengal.

They soon found that no one had been reporting the true extent of smallpox or how it spread. First they would have to understand the problem. Eradication would have to wait.

Nedd Willard, an American public information man at WHO in Geneva, once visited Monnier in Rajasthan. "We went to a rural primary health center (PHC) in an area where there's really nothing but sand dunes and camels—right out of a travelogue—and we met the young local doctor. We said, 'We hear that there's smallpox in this area.' The doctor said, 'Well, nobody comes to my clinic with smallpox.'

"Then we went out and met an old villager and asked, 'Is there any smallpox around?' He said, 'Yes.' We were approximately fifty yards from the clinic, walking slowly. We went into a hut and, sure enough, you can't miss smallpox, not smallpox in Asia particularly, which is so incredibly virulent. And people there had been dying of it. We went back and told the doctor, and we physically went with him to start doing something.

"It showed villagers aren't as stupid as some think. What was the point of their going to the clinic? Even those totally illiterate camel drivers knew that there was nothing the doctor was going to do for them. All he might do is charge them some money for a treatment that is totally ineffectual, and for painkillers they had on their own. It also showed that public health medicine is not that profitable, otherwise, if he thought it was something for which he had a good treatment he would have gone outside to stir up a little business.

"Certain types of health problems can't be coped with in conventional ways," Willard says.

The four WHO epidemiologists learned that lesson quickly and adapted. Monnier was to become a local legend. The Indians called him "the Lion of Rajasthan." He was unusually old for this kind of work, probably in his fifties. Field epidemiology is a young person's game. But Monnier, despite his fragile appearance and quiet mien, attacked the disease with a vengeance, roaming the desert in a borrowed government Jeep.

When a mother would refuse to touch her dying child Monnier would sit on the ground holding it in his arms. One time at least the child recovered and Monnier became a saint in its village.

With a sparse, mostly nomadic population, Rajasthan was ideal for intelligent smallpox-control procedures and, virtually single-handedly, Monnier brought the disease under control.

"He really got the thing going—this isn't the official line with the, quote close collaboration of the local authorities unquote," one WHO official says.

Grasset had some outside help. From time to time WHO

sent in consultants, including Ladnyi (who had left WHO for a post in Moscow) and CDC's Joel Bremen.

But Grasset's major problem was low morale in India. She was confronted with a deep pessimism that the Indian program would ever work.

"So many people didn't believe in smallpox eradication," she says. "People kept saying to me, I always remember, 'Do you know anything about the ecology of India?' I said, 'No.' 'Well, then, how can you get rid of smallpox in India?'

"We had some of our highest officials in India, our director general of health services, who right to the end believed that our . . . program (of surveillance and containment) was not a good one . . . He never believed in it.

"I never really got cross with him because it's like love. You can't make people have faith in something, and you can't make people love you, it's either there or not.

"So many Europeans and Indians didn't believe it, and a lot of time and energy was wasted because we had to fight against this lack of faith. It was never the amount of smallpox that made me sigh, it was the people who sincerely believed that we were wrong and whom we had to convince."

The program staggered on unsuccessfully despite the Lion of Rajasthan and Grasset's élan.

In San Francisco Dr. Lawrence Brilliant had just received an offer from Warner Brothers to make a movie. Brilliant was decidedly uninterested but Warners had offered to contribute to some of his free clinics.

Brilliant, who studied philosophy and ethics at the University of Michigan and had a medical degree from Detroit's Wayne State University, was much more interested in radical politics than in movies or even medicine.

He was an early member of Students for a Democratic

Society, a leader in the antiwar movement and a hippie active in the "counterculture."

He was in San Francisco doing his internship at Presbyterian Hospital when he and his wife, Elaine, fell in with the Bay Area's hippie community, some of Ken Kesey's Merry Pranksters and the Hog Farm commune. Warner Brothers approached Brilliant for help in making *The Great Medicine Ball Caravan*, a film about a bunch of hippies who rode across America in psychedelic buses, playing rock and folk music. It was 1969 and the Age of Aquarius. Only because Warners promised $10,000 for his clinics did Brilliant agree. He and the rest of the Hog Farm eventually found themselves filming in England.*

There they decided to hold rock concerts to raise money to relieve the suffering caused in 1971 by a devastating cyclone in East Pakistan (now Bangladesh).

The group bought a few buses for the relief jaunt. "It took five months. It was wonderful," he says. "You can't believe how wonderful it was. Unfortunately, because the buses kept breaking down we were one month in Istanbul waiting for a water pump to be fixed. . . . By the time we got to the border of India and Pakistan there had been a war; the cyclone . . . was no longer the main tragedy.

"The main tragedy was the devastation of East Pakistan by West Pakistani troops. The borders were closed. Ten million Bengali refugees were coming into the Indian state of West Bengal."

One of the buses went back, but about forty people stayed on in India. Gradually Brilliant and his wife were drawn into an *ashram* run by a guru, Nim Keroli Baba, known to his followers as the "Maharaj-ji." The experience would

* The film bombed at the box office. The Brilliants never saw it.

transform the pair and alter their lives in an unexpected way. They found their guru's spiritualism enthralling and their life in the *ashram* one of infinite contentment.

One day in New Delhi the Brilliants met Nedd Willard. Williard told them about WHO. Brilliant was intrigued.

"My friends in the counterculture had a more serious commitment to the alleviation of suffering than my friends who went to medical school," Brilliant says, "and I had not met doctors at that time whom I particularly respected a great deal." Brilliant had Willard send him an application to work for WHO.

Brilliant and his wife, who had changed her name to Girija ("Daughter of the Himalayas"), put the application under a picture of the Maharaj-ji at the *ashram.* The guru was visiting one of his other *ashrams* at the time. A few days later he returned. Brilliant says that he never saw the application under the picture.

"How much money do you have," the guru asked Brilliant at their first meeting following his return.

"We've got five hundred dollars," Brilliant told him.

"Five hundred dollars? That's not very much. Certainly you must be having more back in the United States? Think!"

Brilliant was now suspicious. The guru had never asked for money before but this sounded very much like the "United Guru Appeal."

"I've got five hundred dollars back in America," Brilliant finally responded. "We owe a lot of money for my medical school training. I had to take out loans."

"What? You have no money? You are no doctor!" the guru said. ("Just like my mother," Brilliant mused. "You're not a real doctor if you don't have a house in Palm Beach, an office in Manhattan. It's terrifying when your *guru* says this about you.")

Then the Maharaj-ji started joking and chanting in singsong: "You no doctor. You no doctor. You n-o doctor. U n-o doctor. U-N-O doctor. U.N.O. doctor, do you understand?"

Brilliant confessed he had not the slightest idea what his guru was saying.

"Yes," the guru explained, "you will go to the villages and give injections. You will be a United Nations Organization [U.N.O.] doctor and work to eradicate smallpox. Smallpox will be eradicated. This is God's gift to mankind, because of the hard work of dedicated medical scientists."

"And we didn't know diddley-squat about smallpox," Brilliant says. "I had never seen a case of smallpox." But because his guru said he should, he sent the application to Willard. Willard wrote back saying that there were a limited number of jobs in India, but he thanked Brilliant for writing.

"Did you get your job?" the guru asked Brilliant.

"No, Maharaj-ji, there's no job."

"It's all right, you'll get your job. Go see them."

Then began an almost endless series of trips from the *ashram* to New Delhi, each one requiring rides in pedicabs, trains, buses and taxis.

In time Brilliant got to speak with Grasset and with Zdeno Jezek, a native of Prague, Czechoslovakia. Jezek had just arrived from a successful health program in Outer Mongolia (the Mongolian People's Republic) and had been assigned to help the New Delhi office. Brilliant was fascinated.

Jezek, a short man with tight, curly hair, was in his mid-forties. It is said of him that if he were to parachute to the surface of an unknown planet he would have a map drawn in a half hour, have everyone organized in an hour, and set up a health system in two. He is fearless, determined and self-reliant. In his years in Outer Mongolia he spoke only Russian and lived on meat like a Mongol. Now he was in India speaking English and eating vegetable curries like an Indian.

He had been to the most remote and difficult places on earth and had turned epidemiology into a series of astonishing adventures, all in the service of others. Brilliant had met very few like him.

But there was still no job for Brilliant at WHO. He went back to the *ashram*, but now the guru was agitated and insisted on an immediate return to New Delhi. Something apparently was about to happen.

"Each time I went I would take off my white *ashram* pajamas and borrow a suit jacket. I would have to make this change, but I wasn't fooling anybody," Brilliant says. "They all knew how weird I was."

In his jacket and tie he walked in the door of the WHO building. At the same time a tall American was entering the building. It was D. A. Henderson on an inspection trip. Henderson agreed to interview Brilliant, but brushed him off as nice, inexperienced and odd. He did suggest a job in Pakistan.

"You're not going to believe this," Brilliant told Henderson, "but I'm going to have to ask my guru."

"No! India!" said the guru.

At about this time Grasset had something of an inspiration. She needed an administrator and a writer. She and Jezek did not write English as well as they spoke it. Brilliant said that he could write. Grasset decided to hire him.

"I can only tell you that she used to carry a notebook which was 8½ by 8½," Brilliant says, "and in it she would write down different things: 'get the laundry done, get the Shah of Iran to give money. . . .' Whatever was written down in that book you could assume would be done. In all the time we were with her, which was the better part of four years, I never saw anything written down that wasn't successfully crossed off.

"She wrote in her book, 'Hire Brilliant.' "

Henderson, apprised of her intentions, says he wired her

that it was one of the nuttiest things he'd ever heard of, and she would have to take full responsibility. Now she confronted WHO's personnel department also.

"I had nothing to contribute," Brilliant says, except for his knowledge of and love for India. "I had never done public health work. I was straight out of internship. . . ." He admits now that if he had been in charge he wouldn't have hired himself.

"Altogether I made twelve trips from the *ashram* to WHO, and each time something else came up that made it impossible for me to be hired. The Indian government didn't want an American; WHO didn't want to hire anybody with my poor credentials; at twenty-eight I was ten years younger than anybody else in the program. It was in its first phase and only the experts, the professors of epidemiology, were coming in. A security clearance was required." *

But one day the Maharaj-ji called the Brilliants in to see him. "He had been very, very nice to us for two weeks, unbelievably nice, giving us so much love and kindness . . . and then he just said, as if ending a speech, 'OK, it's time for you to go now.' The first thing we thought of was that it was time to leave the *ashram* and go to the house. We went around the corner and there was the postman carrying the telegram from WHO. . . ."

Now the generals were gathering.

The government of India knew that there had to be a radical alteration in its smallpox program. Some technical changes had been made. In 1971, for example, a complete switch was made to freeze-dried vaccine, which stood up better in the terrible climate. The program also was commit-

* Brilliant had an FBI file the size of the Manhattan telephone directory but it concluded that though a critic of the war he was a loyal citizen.

ted completely to using the bifurcated needle instead of the jet injector and the rotary lancet. Changes in the reporting system were attempted.

"They were reporting cases," Grasset remembers, "but it was completely unequal from one state to another. There were some every six months, I think, from Assam.

"My first Sundays I actually used to weep trying to send D. A. [Henderson] the report for the whole of India. You received reports from 21 states and territories a week—to put all of this in order was terrible. Some local people would send me certain figures and then I would get others from the central level."

The only solution was uniform, complete, weekly reporting.

"We told them that each primary health center had to report weekly to its district," Grasset says, "and then each district had to report to its state, and the state to the central government, and the central government to us. We got that going—slowly."

Another problem was that if a district had not detected smallpox it would not report at all. Grasset never knew if the absence of a report meant no smallpox or that the report was missing.

By telegram or telephone she began to chase districts and states that were remiss in reporting.

Meanwhile the Indians themselves were bringing some of their best people into the program, including the country's best known epidemiologist, M. I. D. Sharma.

Sharma, a quiet, pleasant man with the build and commanding presence that make one seem taller than the actuality, was educated at Punjab University and at Johns Hopkins. He was worshiped by the nation's young epidemiologists and health officers. He lent authority and legitimacy to the smallpox program because he was not only head of the National Institute of Communicable Diseases but held a position in the health ministry and was a world-renowned malariologist. Also he had the direct ear of the health minister.

Every young epidemiologist in India worked for Sharma at one time or another. His dedication was legendary. He had not taken a vacation in twenty years. He would go anywhere, no matter how remote. He did much to legitimize the smallpox program.*

"I could report directly to the minister," Sharma says, "and since everybody knew it, that made my life easier. Once they knew the minister was with me. . . ."

Others of India's finest were shifted to the smallpox program: Mahendra Dutta and Mahendra Singh, R. N. Basu, R. R. Arora and C. K. Rao. They held responsible government positions, knew their jobs and had the courage to do them.

Only one link was missing in the chain.

In the early '70s some Indians finally had decided that mass vaccination simply would not work in their country. They had watched the Indonesian experience and knew Bill Foege.

"I knew the Indians who were involved in the program fairly well by then," says Foege, who had by this time returned to Atlanta. "We had a series of conferences, one in Thailand, one in India. I was on various WHO committees with people from India, so that I already had fairly good personal contacts."

Foege was not shy about telling the Indians that their mass-vaccination system was useless. It was finally agreed that if there was to be a shift to surveillance and containment, Foege would be brought in. He was more than willing. He was, by now, very restless behind his desk in Atlanta.

---

* Sharma is now retired and lives in Delhi.

"It was clear that there was one major problem in the world and that was the subcontinent. I asked Dave Sencer if I could take a leave and go there." Sencer, head of CDC, strongly supported the smallpox program, and no matter how much he wanted Foege in Atlanta, he knew he would be more valuable in India. Foege joined Brilliant, Jezek and Grasset on the WHO high command.

All the generals were in place and the war against Sitala could begin.

XIV

The War Against Sitala

When 1973 opened there were only four countries in the world with serious endemic smallpox: India, Bangladesh, Pakistan and Ethiopia. Cases were reported in Botswana in southern Africa, but the number was small and the situation was considered easily controllable. Nepal reported smallpox as well, but all cases were imported from India.

Candau had retired as director-general of WHO and was replaced by a Dane, Halfdan T. Mahler, who was more sympathetic.

The regional director of WHO, who had been an obstruction, was replaced by a physician from Sri Lanka (formerly Ceylon), V. T. H. Gunaratne. Sri Lanka had eradicated smallpox and Gunaratne was convinced that it could be done in the neighboring subcontinent. "He became the most aggressive regional director," D. A. Henderson says.

Funds were limited, so WHO initially had decided to declare Indonesia and Afghanistan as priorities while keeping WHO support in India and Nepal at a maintenance level. The last case in Indonesia was in June, 1972, and two years later WHO was forming a commission to study the program in Indonesia. Under WHO guidelines, two years have to pass after the last case before an international commission could be sent to certify that the country was smallpox free.

The day when that would happen in India must have seemed far in the future in June, 1973, when WHO, the central government and representatives of seven states met to discuss the expanded program for India. (At about this time the last case was reported in Afghanistan.) Nicole Grasset represented the New Delhi office. Henderson flew in from Geneva with his second-in-command, Isao Arita, of Japan.

The principal endemic areas were four states, Bihar, Uttar Pradesh, Madhya Pradesh and West Bengal. (Assam later replaced Madhya Pradesh on the list.) An epidemic, the largest in a decade, had swept through Calcutta and spread to areas far beyond that metropolitan area.

In the nation's capital smallpox was still endemic, and new cases were being imported from other parts of India all the time. One outbreak even struck a New Delhi leper home. Fortunately none of the lepers themselves came down with the disease (which would have been an unspeakable horror) but members of their family did. It was only a few miles from the ministry of health.

An investigation into three of the New Delhi outbreaks that year blames "delayed detection, inadequate isolation and containment action." The report said that two months after the initial case, the disease was still spreading in one of the outbreaks. If this could happen in the capital, what would it be like elsewhere?

India's constantly shifting, milling population percolated the disease across the country. People incubating smallpox would travel a thousand miles, leaving behind unseen clouds of deadly viruses. Clinical cases would ride the jammed trains with scarves wrapped around their heads to hide their smallpox from the other passengers.

Of India's 397 districts, 211 reported smallpox. In fact, 50 or 60 districts were responsible for 95 percent of all the cases in India and the majority in the world.

At the meeting between the WHO high command and the government and state health officials, it was decided to concentrate on the worst districts and to take advantage of the seasonal fluctuation. As in Africa, smallpox in India was most widespread in spring and early summer, hit a low point in late summer and early autumn, and then began to rise toward the annual high. The period of lowest incidence was linked generally to the heavy monsoon rains.

Several explanations were ventured for this annual variation. During the rains social intercourse is at a minimum, with weddings and other pleasant events delayed until the fall. About the only occasions that draw crowds are funerals. Another possible factor is that the virus has trouble surviving in hot, humid air.

It was believed that if the program could take advantage of the lull to contain the smaller number of outbreaks, the disease might be eradicated quickly, as in Africa. Some even harbored the hope that the task might be accomplished by December. Instead the expanded program in 1973 demonstrated how much trouble it was in.

Fifty teams of epidemiologists spread out across the four endemic states. Two epidemiologists, Mike Lane and Ed Brink, were sent to Bihar's poorest sections. Lane, one of the authors of the paper on surveillance and containment, found himself in Santal Parganas, named for the primitive Santhali, something of an anomaly even in India's diverse culture.

"They have no health services," Lane says. "They are out

of the system. When I got there eight cases had been reported for the last few months, but in about ten days my wife and I found 800 cases. By the time we left the authorities were reporting about 1,000 cases every two weeks. Nearly one out of every three villages was infected. The population of Santal Parganas was about three million."

Brink in Bhagalapur was finding similar horrors. Side trips into other areas also turned up smallpox. WHO in New Delhi thought that Uttar Pradesh to the west was the main focus of the disease, but it was now clear that Bihar was in the middle of a terrifying epidemic.

"Our immediate recommendation was 'Help!' " Lane says.

Lane had discovered what was basically wrong with the program and why half a billion doses of Russian vaccine largely had been wasted. It went beyond bad reporting, at least in his section of the country.

"The medical structure was not only *not* effective in smallpox, it was counterproductive. The medical officers didn't get out of their clinics and into the villages, and the people who were the most useful and helpful in combating smallpox were the civil authorities, who tended to be more interested in what was really going on in the population at the village level," Lane says. "They had better communication with the villages through agricultural extension workers, police, and the like. So we came to rely more and more on the civil authorities and less and less on medical people.

"Besides, we had been told time and time again by Indian officials that the villages in our areas had smallpox because the inhabitants were ignorant and refused vaccination. . . . We found time and time again, as everyone had who had worked there, that this assertion simply wasn't true. The people didn't resist vaccination, they resisted the *vaccinators.* The vaccinators were members of the Congress party, they were of Brahman caste, they were hostile toward villagers who either were not Hindus or were of lower caste. They came in with a vicious, undiplomatic attitude and were physically abusive."

Sometimes the vaccinators broke into the first house they came to in a village, grabbed a child, wrestled him to the ground and held him to be vaccinated forcibly. Adults who tried to run away were beaten.

"We came in," Lane says, "and simply explained what we were doing, what vaccination was. We sat down and drank tea and tried to treat them as equals. We never found any real resistance to vaccination. When we went into Bihar, the number of recorded cases in the state was less than 1,000; it was over 100,000 by the time we had left."

The smallpox staff in Uttar Pradesh found similar problems. One week 354 cases had been reported in Uttar Pradesh. Simply by going from village to village a week later in a limited area they found nearly 6,800 afflicted. The cases just weren't being reported.

"The reporting efficiency couldn't have been more than 1 or 2 percent," Foege says. "It doesn't matter so much how much underreporting there is so long as you know what that number is." But in Uttar Pradesh the eradicators had no idea what the multiplier was.

The Indians were not only missing cases, they were sometimes deliberately hiding them. If a PHC official did report smallpox to the district medical officer, the district officer would descend on his subordinate demanding to know why the people weren't vaccinated. Then the central level people would descend on the district, raging about the sloppy work at the PHC. It was easier simply not to report cases.

"They would not admit that they ever had smallpox," says WHO administrative officer John Wickett. "You weren't supposed to have smallpox. Government decrees would come down: There will be no smallpox! And these guys just would not report.

"In India there's incredible pressure on the individual. A guy knows that there's a million other guys—literally—waiting behind him for his job. So there's this tremendous tendency not to make waves. You've got to sit there and overcome this inertia . . . to convince these guys that . . . believe it or not, when it comes to smallpox, if you don't do something you're dead. You'd better make waves; anything is better than nothing!"

There were even instances in which people reporting cases were threatened with transfer or dismissal. One day the Ukrainian, Moukhopad, came to Grasset complaining that the Indians were going to throw out his best Indian medical officers because they had been reporting smallpox. Their superiors interpreted this to mean that they had done a bad job of vaccinating before the epidemics. Grasset had to go to the health ministry to save their positions.

WHO decided that one way to solve the problem of nonreporting might be to issue rewards for reporting cases. The central government lacked the necessary authority, but it did ask the states, beginning with those where the disease was not endemic, to offer a reward for each case reported. The states agreed. The rewards were set at Rs10 or Rs25, not an insignificant sum to many Indians.

The government stoutly resisted any rewards to the health workers, however: It was the duty of the health worker to report cases and there was no reason to reward them for doing their job. This sounded fine in principle but was completely unrealistic. Finally, at WHO's urging, the health workers became eligible and a number were encouraged to go over the heads of their superiors and report cases, while WHO kept their identity a secret.

The reward gave official notice that government, at least at the highest levels, wanted cases reported. That assurance probably encouraged some reporting and also led people to track down rumors. There might be a rumor of a smallpox case on a mountain. Without the reward, it was possible that no one would bother to check. With it, there was always someone willing to make the effort.

But it was still necessary to seek out the cases on an organized basis, so WHO and the government instituted village-by-village searches.

"We would go to one house to the east, one to the west, one to the north, one to the south, to the teashops, the market, all the major places," Grasset says. "When we saw that we were still missing some cases, we said, 'All right, let's do a house-to-house search.' Again, people just never believed we could do this. They said, 'How can you do a house-to-house search? It's impossible, and especially in the towns.' In towns, of course, you could go floor by floor to see how many families there were. Finally, when you got down to it, even room to room. I said to the skeptical, 'Remember how you said it was impossible even village to village?' "

One problem was that the eradicators did not know where all the villagers were nor, as in Africa, the location of the satellite hamlets. Village and hamlet populations were unknown and no one had reliable maps.

Grasset requested every epidemiologist who passed through her office to treat mapmaking as the first priority. As soon as the epidemiologist got to his assigned territory he was to sit down with the local health workers and plot all the main roads in the area. Grasset told the epidemiologists that the maps should indicate the location of all villages and hamlets and their populations, at least approximately. Eventually each epidemiologist had data listing every household in an outbreak area; the number with smallpox; who had been vaccinated.

If there was an outbreak, everybody in the thirty or forty houses close to the infected family would be vaccinated. Soon the population to be vaccinated was increased to all within a five-kilometer radius of the infection and a search for other cases was instituted for an additional five kilometers. A follow-up search two weeks later would catch anyone who might have been incubating the disease. Any newly discovered cases would be the focus for a repetition of the same procedures.

Victims were isolated in their homes. Guards were hired to sit in front of the door to keep everyone not vaccinated out and to prevent the sick from leaving. "Hospitalization of smallpox cases was completely stopped," explains Sharma, "because experience showed the spread of the disease was more from hospitals. The isolation was carried out in the house. It was convenient for the family as well. Whatever (medicine) was needed was given."

Home isolation had another advantage: the isolated families received subsistence pay. Previously they had been terrified of being sent to infectious disease hospitals and having their livelihood disrupted, and so they would hide cases.

The guard system required expansion. It was quickly discovered that having a guard for each door in the house did not work. If the guard fell asleep or went to the toilet the victim might go out to visit neighbors or the central water well. So two guards were hired for each door, one to relieve the other. Four might be needed. Isolation lasted as long as six weeks.

Beginning in October, 1973, surveillance and containment was instituted as the method of eradication. Every health worker in India was sent out for one week out of every four or six for the village-by-village, house-by-house search. Eventually a hundred thousand were involved in the program: thirty thousand worked full time on smallpox, the remainder were borrowed from other health programs for the mass searches.

As the operation grew the eleven people running the program developed a very special relationship which they felt was largely responsible for its eventual success. The war board consisted of four permanent internationals: Larry Brilliant, Foege, Jezek and Grasset. A fifth was one of several administrative officers from CDC who rotated in and out every few months. On the Indian side—the government's central assessment team—there were Sharma, Rao, Basu, Dutta Arora and Mahendra Singh. One international and one Indian were matched per state or area. Jezek and Basu, for instance, might be teamed in southern India, Dutta and Brilliant in Bihar, Foege and Rau in Uttar Pradesh. The assignments changed occasionally. What was important was the deep abiding respect and intimacy that sprung up among the principal eradicators.

"We really developed an esprit de corps," Brilliant remembers, "and a camaraderie which makes that group of ten people still very close friends. There wasn't a bad apple among us. The task brings out something in you you didn't always know you had. And I think each person will say that he worked harder and better and more honestly and more honorably in the quest for the smallpox eradication than in anything before or since."

Foege says, "It always remained informal, an excellent working relationship. It may be hard to grasp from the outside, it may also be hard to replicate, yet I'm convinced that this was the sort of thing that made the difference. There were no longer WHO recommendations or Indian government recommendations that had to be cleared by the other side. . . . We would sit down with each other and we had an objective and we could reason it out and get there . . .

"If you are assigned to a country and you go in, no matter how many times a week—two, three or four—and sit down with someone, that's an altogether different thing. It can never really lose its formality," Foege says. "It's altogether different from traveling back together and continuing to discuss what happened at the meeting, what should have been done differently. You become part of a family in a way that rarely happens but that's the key to having things work or not."

Grasset cites an Indian philosopher who says that real happiness is to have only one aim in life, one so high, so great, that it eliminates all else. "That's what we were all doing. Having a common ideal made us coordinate our efforts better."

After the initial shock of the early autumn outbreaks, the WHO team in New Delhi called for reinforcements. Geneva WHO responded with a small army of young, tough epidemiologists from all around the world. Most of them signed up for two- or three-month tours of duty. They spread out across northern India (a country few of them had ever seen before). It became one of the largest international campaigns of its kind in the history of medicine.

WHO's Geneva office soon found itself running a circus, a jubilee of unorthodox medicine. Nothing like it had ever happened before to the stodgy old bureaucrats by Lake Leman.

"They were the most unconventional, the most un-WHO types," said Willard. "They all looked like sloppy American college students. They were not all Americans. The Russians, Swedes and French were just as sloppy as the Americans. But they don't look like the usual international civil servant with tie, shirt and jacket. They were field-oriented. People who like being in the field, like being in a Jeep, hated putting on ties. It's hippie medicine. It reinforced itself. The sort of person who doesn't like putting on the tie is the very sort of person who likes going out into the bush in a Jeep and vice versa.

"The fellow who is looking for the office job—and I've known cases, even with CDC people in Atlanta who don't like that field stuff so much—it's amazing how quickly they gravitate to an office.

"It was a very young campaign," Willard goes on. "That's another way you could recognize smallpox people. Their average age was so much lower. Most of our people in WHO are postgraduates, which means most of our people are post-forty. Most of these kids were from about the twenties to early thirties. It was a very young group."

The people and the situation demanded innovation and flexibility and courage. Rule and precedent fell everywhere. Instead of sitting in an office and ordering the natives about, the army of young epidemiologists roared off in Jeeps, Land-Rovers, locally built four-wheel-drive vehicles, on motor bikes, in rickshaws or on foot. Everything about them was unorthodox.

"Sometimes they were living with women who weren't their legal wives. For WHO that's incredible," Willard says, "because then you're not an official dependent, you don't have the right status. I must even mention that some of them were Americans. But, to set the record straight, one was a Brazilian. And another, a Frenchman, lived with someone from the country itself. Shock, horror, dismay!"

The bureaucracy in Geneva could not understand that the medical and operational innovations and rule breaking that they considered unsuitable, immoral or simply wrong seemed absolutely necessary to the people in the field in India. The Indian staff repeatedly told Geneva, if you have not been to India you cannot understand. The guards were one matter. It was against the rules for WHO to pay for them. Where was the precedent?

In one instance a WHO worker learned that outbreak reports were not showing up from an area because the local health officer did not have money for the postage. The man from WHO told headquarters, "We've got to provide these people with money for stamps."

"The real bureaucratic types said, 'I beg your pardon, Rule Number 32/AB130 for example says that under no condition will WHO provide money for office supplies to a national government,' " Willard relates. "If it hadn't been the smallpox campaign that would have ended it. What did they do? Very simple, the next time the guy went out he had an envelope. In the envelope he had cash and the cash was listed for 'operating expenses for the Jeep.' " The "operating expenses for the Jeep" were converted into postage stamps and the reports resumed. (Willard says that after the smallpox program ended the Indian official again ran out of money for stamps and no further reports were mailed.)

"The idea of a reward sounds like a bounty, sounds crass and craven," Brilliant explains. "For WHO, that kind of money for citizens to do something that seems like a good citizen's duty outside of an Indian context, that's absurd." Grasset supported her staff in paying rewards because they were in fact necessary.

Grasset, too, would spend much of her time fighting with the traditional bureaucracy in Geneva but mostly at WHO in New Delhi.

"A lot of the hierarchy didn't like—still don't like—the way the smallpox people ran their program," Willard says.

"The rules and regulations we had to break!" Grasset exclaims. "We had to get it officialized. You can't go on breaking regulations, they have to be officially broken."

One example had to do with office supplies flown in to India from outside. From the moment the shipment arrived at the airport, the Indian government was supposed to be responsible for delivering it to the right place. But it was not always able to cope, and, against WHO regulations, the WHO Indian staff sometimes had to take over the internal movement of the supplies. There was also a rule against WHO's paying for printing; Grasset had to get special permission to take over the responsibility for getting the forms and other material printed. In violation of regulations,

WHO sometimes took over delivering vaccination kits that otherwise might stay undelivered for a whole year.

There was also the Indian bureaucracy to contend with. Perhaps there is no place on earth with a bureaucracy quite like India's. Nothing can be done in the country that does not involve dozens, sometimes hundreds, of people, all with ball-point pens and rubber stamps.

Sharma acted as a buffer between the Indian bureaucrats and the program. "We developed a system of level jumping," he says. "You have to go through certain levels. You see your boss, your boss sees his boss, and so on. But level jumping means that you don't bother with the two fellows just above you. Go to the top. We level jumped and I could go to the minister of health, Karin Singh, whenever I felt like it. I would ring him up, or he would ring me, any time."

That helped because Singh usually was completely supportive of the program. He absorbed many of the pressures from elsewhere in the government, protecting even Sharma.

The people in the field soon adapted an "us-them" attitude toward the WHO and the Indian bureaucracies, but any blanket condemnation would be unjust. Many WHO bureaucrats both in Geneva and in the field worked superbly, with talent and courage, and the program would have failed without their initiative and support. There were administrators such as Jack Copland and Ron Hauge, both Americans.

One such bureaucrat was John Wickett. Wickett, a slim Canadian with a thin, neatly manicured beard, had arrived in Geneva in the early seventies with the clothes on his back and a guitar, hoping to find a job picking grapes. Somehow instead he managed to get a job with WHO as an administrative officer. He married another Canadian WHO employee and became an operations officer in the field.

It was Wickett's responsibility to provide supplies. If a Jeep was needed, Wickett provided it. Sometimes it was better not to ask where the vehicle came from. He probably never stole anything, but there are confirmed cases of WHO administrators "borrowing" vehicles.

Many of the administrative officers became geniuses at improvisation. When a bureaucratic snafu in Bangladesh prevented vaccine from reaching a refugee camp in India where it was desperately needed, one WHO administrator simply smuggled it across the border.

They also became genuises at playing the political games. They learned quickly that direct confrontation did not work in the game of office politics. Many learned to love the game.

"You really have to keep your cool," one of them advises. "You have to bring everything to bear on your problem."

Nothing at WHO (and other U.N. agencies for that matter) is not highly political. The U.N. requirement of "geographic distribution" means that its agencies' employees have to come from the broadest possible spectrum of countries. As a result genuinely mediocre people from the right country get jobs in preference to others far more qualified.

"You have to find out where your real channels are," Wickett says, "and operate through them, I mean find out who really does the work.

"You play the game, otherwise you go crazy. But it's fascinating because it's such a challenge to find out how to get it done. There's always a way. You just have to be a little more tenacious than everybody else. You can wear anybody down."

Wickett played the game well. When a vaccinator or a WHO worker in the field needed something in a hurry, Wickett or other administrators like him would get it quickly despite obstacles. There were a number of bureaucrats like him in WHO and they made it all work.

"In public health," Wickett explains, "everything is an emergency."

There were other, more personal, ways to fight the bureaucratic wars. Henderson's, for example:

"D. A. was never one to discount the fact that he was a tall fellow," Wickett says, "and he had a nice, rich voice and no hesitation whatsoever to browbeat people, no hesitation at all. . . . I remember one time, I thought the walls would blow out of the office. People at WHO weren't used to such undiplomatic behavior."

The Indian bureaucracy was no less capable of turning out innovative and creative people when they were needed. "Many of the Indians rose to the call of duty in a way that makes them more heroes than we were," Larry Brilliant says, "because we always could go into that phone booth and put on our Superman costume and become white, wealthy and have a ticket home. They didn't have that option. They were home. They were going to be there and face all the superiors that they disobeyed after the smallpox program left."

The year 1974 was to be the low point of the campaign as the eradicators began to suffer from success. Their surveillance system became so good that it was finding and reporting more cases than ever before in history. On top of that, Bihar was hit by floods and relief resources had to be diverted there. An epidemic exploded and soon was exported to other states as migrant workers fanned out from the factories and construction sites to their homes elsewhere.

Weeklong searches continued in Bihar. The teams consisted of a vaccinator, a paramedical assistant (PMA) and a driver, sometimes bolstered by an international (i.e., foreign) or an Indian epidemiologist. They continued their isolation of discovered cases, building a ten-kilometer barrier of the vaccinated around every case.

By the arrival of that summer's monsoons in Bihar alone there were forty-five international and more than a hundred Indian epidemiologists. Over a hundred fifty containment teams were roaming the state. Twenty-six epidemiologists from overseas were sent to Uttar Pradesh.

No one knows if there really was more smallpox in 1974, but since the improved surveillance system found more cases it appeared that the disease was increasing. The diehard believers in mass vaccination bubbled to the surface again and revived the old fears of surveillance and containment.

"In 1974, up until June, I was under great stress and strain," says Basu, who worked at the health ministry. "Whenever they asked me, 'What is the progress of the program?' I used to tell them the number of cases that month. It was always higher than the previous month. And they used to laugh at me and make sarcastic remarks: 'Dr. Basu says that the progress is satisfactory because more of our people are dying.' And there was opposition to this campaign. Our ministers had to face questions in the national parliament. Luckily Karin Singh was very enlightened. And that time we got the moral support of WHO. I used to bring Dr. Foege and introduce him to the administrators to assure them that surveillance and containment was the right procedure."

But the Bihar state government panicked. The industrial complexes of Dhanbad and Singhbhum in the southeast were particularly hard hit. The state government insisted on reinstating mass vaccination and taking its health workers away from the surveillance-and-containment teams. It insisted that its approach could eradicate smallpox in six weeks.

The Bihar minister of health went to New Delhi to demand assistance. Basu begged for two months more.

"We have the feeling that it will take two months," he said. If there was a decrease in the reported cases within that time Basu was convinced that the state government would be satisfied and would not disrupt the program. The Bihar health minister agreed, but the state minister did not. He went to New Delhi and at a meeting of epidemiologists told them of his plan to swamp the state with vaccinators. He asked the others their opinions.

One epidemiologist stood up and said, "Sir, if a house is on fire what will you do? Will you take the small quantity of water you have available and try to stop the fire in all the village, or will you go to the house on fire and use all the water there?"

With limited resources, WHO and the central government could not support both mass vaccination and surveillance and containment. The state government relented and left the program staff alone for a while.

"We were not delivering the goods, and he suspected us," Basu explains.

The central government was not unified either. The director-general of health services who served under Karin Singh never believed in the new strategy. He sent a letter to the program officer in western Bihar telling him to be careful of the uninfected towns and to see that their people were vaccinated immediately. The officer obeyed and surveillance teams were dismantled and sent to places where there was no disease. Grasset and her staff raced to the health ministry. Finally the director relented, but he was so furious that Grasset was sure that she would be kicked out of India.

Grasset and D. A. Henderson had been a little concerned with all the new epidemiologists. The days when they had handpicked their staffs were over. It turned out that only a few of the newcomers had to be sent home or shirked their tasks.

"It wasn't always the men with the most experience who did the best jobs," Grasset says. For example, one man, a high-ranking health officer in his own country, was sent to Bihar and worked out a marvelous plan, but the frustrations of working in India were too much for him. He left in three months with more cases than when he had arrived.

"The last thing he said to me was, sarcastically, 'Well, Nicole, have a good time in India.' That was a man who never believed in the surveillance method. I was cross with him but not discouraged. He wrote Foege a letter telling him what rubbish surveillance was."

Life in the field was not easy, which discouraged some. Occasionally an epidemiologist would be lucky and be able to spend the night in a government inspection house built by the British in rural areas.

"You could cook your own food," says CDC's Walt Orenstein. "There was always *samosa,* which is essentially like a knish. It was potatoes with chilis wrapped in dough and fried in hot oil. It was all starch but when you took it out of the hot oil you could eat it anywhere. There was a lot of fruits and peanuts. At first I was shocked at the type of places I had to eat at, but I never got sick in India."

Sometimes the epidemiologists would spend the night in the villages. The huts were made of brick held together by dried dung but were fairly substantial. Most of the towns were basically a square around the communal well. Animals were everywhere, especially chickens and goats. The occasional stray cow would wander by, adding its droppings to the filth of the other animals and of the village's children. Most of the houses had several rooms and a front porch. Furniture was sparse. Most people slept on cots or on the ground.

The traveling epidemiologist every day would search four or five villages with an average population of about a thousand.

"They had forms listing the villages and other data," Orenstein says. "The form gave the village name and the day it was supposed to be reached and the names of the vaccinator and other workers. Each worker got a shorter form. Workers often went to the marketplace first, trying to get people together. Each worker had his own technique. Some were quiet and shy and went from stall to stall. Others tried to collect a crowd. After visiting the marketplace they were supposed to go to different areas of the village. Often they would talk to the village head, the *pradhan.* Usually he knew what was happening. Later, as we got more sophisticated, we wanted the workers to go door to door."

If workers found a case, guards were immediately posted at the doors of the infected house. Then, going in concentric rings around the victim's house, the workers took a village census, marking down the vaccination status of each person. The houses were numbered in chalk or crayon on the wall to direct each vaccinator to the twenty or thirty houses he was responsible for.

Everyone within a five-kilometer radius of the victim's house was vaccinated. Everyone within ten kilometers was questioned for cases. Orenstein soon developed a street theater ceremony to accompany bestowal of the reward for reporting cases. He made sure that everyone was called to the center of town to witness handing over the reward money, by this time up to Rs100. Orenstein then had himself vaccinated so that everyone could see how painless it was. He was vaccinated dozens of times during his stay in India.

The program attracted a number of free-lance adventurers like Alan Schnur. Schnur, who had left college to become a sportswriter in New York, somehow found his way to WHO and became a professional eradicator. He worked most of the year on short-term contracts and vacationed with his girlfriend the remainder of the time on the money banked for him in Switzerland.

Schnur, a tall, almost scrawny man, capable of going anywhere and traveling alone, generally dressed like a native, living off the land or the local hospitality, carrying only a sleeping bag, a roll of toilet paper, some WHO forms and a bag of vaccine and needles.

He acquired adventures the way most men acquire haircuts. He could spend months in the field and sometimes would go where even the natives feared to tread.

Schnur and a few others would turn up whenever WHO needed people who could disappear into the bush to vaccinate the inaccessible. They would prove their merit further in East Africa.

Surveillance was still the *sine qua non* of the program, and to find cases various schemes had to be devised. New graves were checked to see if the body was a smallpox victim. Flowers on a shrine to Sitala led searchers to scour the neighborhood for possible smallpox nearby. Recognition cards showing pictures of infected children were flashed around the markets, tea shops and village wells.

"Case finding is the fun part of epidemiology," says Mike Lane, "that's the detective work and the medical anthropology."

Sometimes even getting around was part of the fun. Mary Guinan, a tall New Yorker sent to Uttar Pradesh from CDC, once found herself blocked by a river from returning late at night to her headquarters in Lucknow.

"The river is now unpassable," she says. "It's about neck-high. How am I going to cross here? I can't possibly stay here tonight. We didn't have any of our supplies. There was no food. There was *no* food! And it was freezing, it was 40 degrees. They finally rigged up a system with two villagers carrying a bicycle on their shoulders and I got on top of the bicycle. And these two guys walk across in neck-high water and carried me on to the other side."

But that was the hard way to cross the river. The easy way is to take an elephant.

After a week or so she met the local landowner, the *raj-sahib.*

"He could speak English. He lived in some remote area in a huge mansion that was falling apart, but he owned all of the surrounding lands. He came to ask me if he should get a smallpox vaccination, and I told him that he should. And then he offered me the use of his elephant. An elephant is a sign of wealth in India. If you can feed an elephant you're rich. I think that's why he kept it, to prove his wealth. It wasn't really terribly useful. But for me the elephant was terrific! It came with a driver."

Elephants can swim. It was trained so that if the water got deep enough for the person on its back to get wet, it would pick him up in its trunk and swim to the other side. "It was incredible," she says. "It never happened to me but it did to some of the workers.

"I had lots of things to carry me across the river. Sometimes I could use a camel, sometimes I could get a boat. But if I knew there was critical timing involved, . . . I would try to arrange in advance for the elephant. The elephant rides pretty smoothly and it's pretty tall. Usually I got on with a ladder, but you could step on its trunk and it would put you on its back. I was always a little nervous about that."

Mary Guinan used elephant commuting to control one of the largest smallpox outbreaks.

Except for monthly meetings in which all the internationals would get together, they were always with their Indian compatriots in the field.

"You spend much of your time traveling and repeating instructions over and over again," says Guinan, who was known in Lucknow as *"docmemsahib."*

"We had a team of a driver and the paramedical assistant, who really functioned as an interpreter," Orenstein says. "He had been working in smallpox for several years and he knew how to deal with it. I happened to be very lucky. I had three PMAs in my time, two were excellent, one not very good. But the two who were excellent knew who was corrupt, who was weak, whom you had to grease, and knew how to take care of you.

"Here I was never having seen a case of smallpox and

supposedly a smallpox expert. It really happens. I was in India maybe a day or two and all these guys who had seen smallpox all their lives call me over on a case that's difficult to diagnose and I'd have to say, 'Well what do you think?' If you trusted the PMA you would listen to what he said. He really ran it. In some sense you were there for show. You were his authority. You were his power."

WHO guaranteed the PMA's salary and almost doubled it. Being a PMA was a popular position.

One such PMA was R. P. Singhal, based in Lucknow. He worked out of the smallpox office in the commercial center of the city, in a ramshackle building behind a wall and gate. The unit's Land-Rovers were parked on the patio next to a loading dock. The office, cooled by a ceiling fan, was covered with charts and graphs chronicling the outbreaks and the containment procedures.

Singhal, a hearty, intense man then in his mid-thirties, worked with eight or nine different epidemiologists in the course of the program and probably vaccinated more than 10,000 people. He began fighting smallpox in 1962.

Singhal was specifically trained at the Lucknow headquarters by program officers for three months: His education included college but in the school system of India this was the equivalent of a junior-college education. Indians spend less time in secondary school than Americans. He took courses in economics, political science and English literature but no medicine or science. That was added to the curriculum later in the sixties. PMA training included vaccination, the study of the vaccine and the elements of epidemiology. Much of it was field training.

Singhal found that there was very little resistance to vaccination in villages where the disease was raging. He would point out to the resisters that the sick children had never been vaccinated while those with vaccination scars did not have smallpox. That was easy enough for anyone to understand.

Because children receiving primary vaccinations got sick about seventy-two hours later, many parents tried to avoid the vaccinators. The vaccinators would try to leave the village before the vaccination fever began.

In villages where there was no disease, resistance to vaccination was stronger. A few times Singhal had to cordon off a village and use force to vaccinate.

For more than ten years Singhal went from village to village. During the mass-vaccination era he aimed for quantity. When surveillance and containment came in he became a principal cog in the new strategy.

The PMA's work was closely supervised by medical officers such as R. S. Bajpai, also of Lucknow. A pleasant, mustachioed physician, Bajpai could make considerably more money in outside practice but stayed with the Uttar Pradesh ministry of health and the smallpox program.

"Supervision was almost constant," Bajpai says. "Somebody was visiting every day, either the WHO epidemiologist, or the district medical officer, or the chief medical officer. The vaccinators must be in infected villages for four weeks. We had three or four vaccinators for each village, plus a supervisor, who was in the area for six weeks, so supervision was very close. He would know what vaccinator was working, which was not."

Eighty to ninety percent of the cases could be traced back to their source, Bajpai says, if enough time and energy was expended.

"The monthly search was organized by our own staff," Bajpai says, "which included health workers, welfare workers, and the maternity workers. Nobody was given an exception. In every primary health center there were about twenty-five to thirty workers. There are about a hundred villages in each PHC, so we gave four villages roughly to each worker. In these four villages the workers have to search six days of every month. Later on we decided to have independent searches to make sure that the villages were being searched by the original teams." Bajpai hired high school students to do the back-up searches. They were given Rs5 a day and excused from school. The PHC people were really doing the job, the checking showed. (Other health programs suffered because of the drain of manpower to the monthly searches, Bajpai says.)

Vaccinators, like Bagamber Singh, were the backbone of the efforts. Singh, a small, tough wiry man with intense eyes, spent almost all of his time in the field, rarely seeing his family and sending messages back to them through the medical officer. During most of the month he roamed from village to village with his canvas bag of needles and vaccine, living off local hospitality. When the extended searches were under way he would be part of a mobile team based in Lucknow that would be thrown into villages where outbreaks were spotted.

Like most vaccinators Singh had a ninth-grade education. He was given twenty days of training in how to vaccinate and to make the differential diagnosis between smallpox, measles and chickenpox. He spent anywhere from ten to thirty minutes with each family to convince them to be vaccinated, he says. He carried about six packets of needles and vaccine, each packet good for more than twenty-five vaccinations. Drivers and PMAs coming and going from headquarters would keep him supplied.

"As regards personal comfort, I did not have very much," he says. "The village people, they themselves provided it, the bedding, for instance." He could be out as long as a month without a break. When he was not vaccinating he would sit around a fire or in the tea shops talking with people and encouraging them to help him vaccinate the reluctant.

"I have never been beaten but I have been abused many times," he says. "Very bad language."

Other vaccinators, however, were beaten, some quite badly. One man was attacked and stripped naked by a band of *dacoits* (roving outlaws). He went right back into the village, guarded by friends with shotguns.

The cities were even more difficult to handle than the villages, Orenstein says. The people were much more sophisticated and more suspicious of the government. Sometimes the eradicators had to use extortion in places like Lucknow and Aligahr. People were told that their food ration cards would be taken away if they were not vaccinated, and any member of the family whose name appeared on the card also had to be vaccinated or his name would be erased. Many families relied on these cards for survival. Others were told that if they were vaccinated they could add people's names to their cards. It was compulsion of the crassest sort but it was necessary.

Unforeseen problems were always springing up in the field. For instance, there were India's legions of beggars, some of them professionals.* Some would go from village to village, begging and spreading disease. One man was found to have infected thirty-five people in twelve villages. The beggars refused to be isolated because they lost their income if they were not out on the streets.

The solution to that problem was simple: Support them.

It began when one CDC epidemiologist, Steven Jones, sent Grasset a bill for Rs1,600 for the hiring of a tent, rice and a servant to look after some beggars for six weeks isolation. Grasset was with Foege when the bill came in.

---

* One American researcher passing through New Delhi once watched a child "work the crowd" near a market area. He found that the child made more money in a day than the average Indian laborer in a week.

"I can see myself give this bill for Rs1,600 to our finance officer," she told him. "He'll go mad."

"The next day," Grasset says, "one of the very few Americans who had to be sent back wrote to us: 'You know, I've had a double salary from WHO and I have been paid in the States. I'll give you back your WHO money and you use it to feed the epidemiologist, because it's so difficult in the field to eat properly.' So we wrote back and asked: 'Look, would you mind switching the money over and feeding the beggars, because that's what we need now?' In the long run this is what we did. We isolated and fed all the beggars and all the nomads and homeless persons who were exposed to smallpox."

For a while WHO was supporting a small part of the beggar population of India.

Beggars were a problem for another reason. In places like Calcutta street people—those who live on the sidewalk—had to be watched and vaccinated but no one knew how many there were or to what extent they moved about. A young Canadian woman, Beverly Spring, questioned a sample of the beggar population of Calcutta and assured WHO that there were fewer of them than was thought and that they did not wander around so much as had been feared. It was the first such study ever done of this large group of people.

The number of Indian heroes is beyond counting, Larry Brilliant says. A. G. Achari in Bihar, "the most honest person on the planet," was the deputy director for health services for smallpox in the state during the height of the epidemic. "This man, at tremendous personal costs, disregarded his entire family life for two years. He worked thirty days a month, walked hundreds of miles."

One Indian official suffering from asthma wouldn't leave the coal-mining area he was assigned to even though the dust in the air was killing him.

For counsel the WHO general staff relied on everything

from the *I Ching* and Brilliant's guru. "Maharaj-ji was never wrong," he insists.

And sometimes, the personal toll was enormous.

During that year Grasset suffered from a kidney stone.

"She'd come to work at 3 o'clock in the morning and leave at midnight," Brilliant says, "but she'd rest on a hot water bottle to stop the pain." The surveillance technique was finding unprecedented numbers of cases and the international news media was calling it the worst epidemic in history. As that furor raged above their heads, Grasset suffered her worst attack of pain. Only she and Brilliant were in New Delhi at the time.

"She stayed in New Delhi for two weeks with that kidney stone, excruciating pain, to keep this whole thing together, to keep the entire program working.

"I had no idea of all the things she was doing until she had to be evacuated to Europe . . . I had to take her place. . . . then D. A. [Henderson] had to come from Geneva and Bill [Foege] from Bihar to help us out, because without Nicole there was a gap that nobody else could fill," Brilliant says.* "As competent as all the other people were, Nicole was the spark that held it all together. Nicole provided a spirit of decency and honesty and commitment that really set the tone."

It was the time when Brilliant managed to get into trouble by criticizing a local official in the press. He was nearly thrown out of India by the irate prime minister, Indira Gandhi.

As soon as the pain of her stone eased, Grasset was back at work.

---

* Henderson arrived to find Brilliant suffering from pneumonia, Foege with a herpes infection and Jezek with his feet under a severe fungus attack.

Others sometimes found living in India a harrowing experience. Jezek's wife once climbed into a Jeep and was about to drive off when someone screamed. She looked down and saw an infant's arm under the front wheel. The mother had placed it there deliberately. If killed, she hoped to collect a considerable indemnity. If only maimed, the child would make a more effective beggar. Intentionally maiming children was not unheard of, particularly around Bombay.

One Russian accidentally hit an Indian with his car and narrowly avoided being lynched by irate bystanders. Wickett once struck a man with his motorcycle by accident and had to flee lest he too be attacked. At the nearby smallpox office he returned with help and took the slightly injured man to a hospital.

Yet, despite all efforts, the epidemic in Bihar made eradication irrelevant in the early part of that year. Merely controlling the disease was hard enough. In few places on earth was smallpox worse than in the Chotanagpur division, Singhbhum district, of the state of Bihar. The major steel-producing area in India, linked by road and railway to the rest of the country, it was also the world's leading exporter of smallpox. Because there is work and money, particularly in the city of Tatanagar, beggars and farm people from all over the eastern part of the country gravitate there.

The population is about eight hundred thousand, many of whom live in company towns or railway compounds. The administrative setup was chaos, with fifteen autonomous units duplicating services and bureaucracies. No one was in charge.

Searches had turned up more than two thousand cases in just the forty-five kilometers around Tatanagar with its six hundred thousand people. But the local government was not responding. A later investigation showed that officials claimed never to have heard of the outbreak. The local deputy commissioner said that all he knew was what he read in the newspapers. The manager of the largest steel mill claimed that he never knew about the smallpox despite the fact that his employees were among the victims.

The fifty-bed isolation hospital was deluged with more than a hundred cases, so the local high school was turned into a hospital, jammed with suffering children. Hemorrhagic smallpox broke out in the maternity hospital and fifty pregnant women bled to death.

The river that ran through the city was reported clogged with bodies.

Larry Brilliant sought help from Tata Industries, the single largest privately owned company in India and the major influence in the area. The Tata people turned over workers and transportation and were able to slow the epidemic but could not stop the exportation of cases. Tatanagar was completely block off. The railway station, which was probably the single worst source for exporting the disease, received special attention. No one was issued a train ticket unless he showed evidence of a recent vaccination. The thousands who traditionally pushed their way into the third-class cars of Indian trains without tickets found armed guards and barricades blocking their way. Even with this precaution, the homeless who dwelt on the railroad platforms in nearby cities soon became infected and the epidemic kept spreading.

Grasset knew that the only answer was J. R. D. Tata himself, India's biggest industrialist and a seventy-year-old Parsee. Visitors to India ride to their hotels in his Tata Industries buses. The steel and concrete used to build the hotels come from Tata mills. The soap in the bathroom is from Tata chemical works. Tata probably owns the hotel. J. R. D. Tata, as head of the enterprise, was a very busy man.

"I always go by what my mother used to say," Grasset explains. " 'It's better to address yourself to God than to his saints.' " She sent Tata a letter in his Bombay office. A secretary wrote back saying that Tata would be in New Delhi soon and would contact her. A few days later her phone rang.

It was Tata, in a fury because he had gone through three telephone operators to reach her.

"I'm here for one day only," he told her. "I've got meetings the whole day. We've already helped you in Tatanagur, what is this, you want more?"

"Just give me five minutes, that's all I'm asking," she said.

He said that he would give her ten minutes.

She and Brilliant went to see him, nervous because they were really unprepared with formal proposals and had simply written down what they could think of in a hurry. They believed it would cost $500,000 and would take six months.

"We went, the two of us, trembling while wondering what he was going to say. He knew nothing," she says, "he asked about it, he was a bright and good man. After twenty minutes he understood. He said, 'All right, I will help you.' "

He agreed on condition that Grasset get permission from Indira Gandhi. He would not intrude over the local government without the prime minister's approval. She gave it and Tata mobilized his entire company into a smallpox army. What he did not own he bought. If he needed more people, he hired them, including epidemiologists, public health workers, drivers, vaccinators, clerks and nurses.

OXFAM, the British charity, also pitched in with vehicles. SIDA, the Swedish International Development Agency, through its representative J. Tranneus, gave a total of $13 million. Other charitable and civic groups like Rotary and Lions clubs and Hindu religious associations also pitched in.

In the end, the Bihar epidemic was defeated by a combination of all these, WHO, Tata and the monsoons.

The staff took advantage of the break in the spread of

smallpox caused by the monsoons of 1974 to tighten its surveillance-and-containment techniques.

Epidemiologists from around the world continued to pour into India. By the time the campaign ended 236 of them, representing thirty countries, fought in the India campaign. Of that total, sixty-eight were from CDC; there were thirty-one other Americans as well.

When the monsoons were over only Uttar Pradesh, Assam and Bihar still reported any significant amounts of smallpox. Now vaccinators and searchers criss-crossed the three states, responding *en masse* to any outbreak. This was the time when variola was most vulnerable.

In a normal year the number of cases would gradually increase as the autumn turned to winter. But the autumn of 1974 was different. The number of cases steadily declined because the containment measures were breaking up the chains of infection.

By January, 1974, the number of cases in all of India totaled 1,010. The year before the total had been 9,353. February was even better. The 1973 February load had increased to 14,764. But in February of 1975, there were only 212 cases in India. The decrease continued in March, with only 84 known cases as opposed to the 19,554 the year before. By May, which should have been the height of the season, there were only 25 outbreaks, all small.

The last case in Uttar Pradesh was of a little girl, Sanwati, who died in mid-March of 1975 in Aligahr.

On May 5, 1975, a five-year-old girl, Phul Kumari, in the village of Pachera, Bihar, was burned in a home accident. Her mother took her by foot to a homeopathic healer in the bazaar in Barsoi. He noticed that the patient showed fresh pox marks but said nothing. However, a shopkeeper had the mother take the child to the nearby temple while he tracked down a smallpox searcher. The smallpox officer sent the girl back to Pachera.

Alerted smallpox eradicators who had been working in the adjoining state of West Bengal entered Pachera on May 10. They found that there had been some fifteen smallpox cases, including one death. Everyone in the village was vaccinated, but there were two more cases incubating: a seven-year-old boy, Manjo Jogeshwar, and his eighteen-month-old sister, Dano. Both recovered.

The two children were the last indigenous cases of smallpox in the history of India.

There was to be one more victory.

On May 10, the day the eradicators were entering Pachera, a thirty-year-old woman, Saiban Bibi, traveled from her village in Bangladesh by bus and foot to the nearby village of Itauri, where she stayed for four days. There was smallpox in Itauri, she said later, and a few deaths. She had seen the victims along the main street but denied any close contact with them.

On May 21 Saiban Bibi walked across the Indian border and passed through Gobindapur, Fakira, Akhbarpur and a nearby village and proceeded to the train station in Baraigram to sleep on the platform. During the night she developed a high fever. The next day she went to the station at Karimganj and remained shivering on the platform there for three days. She finally moved into the third-class waiting room on May 24. There she lay on a bench, covered with sores and burning with fever. The only time she got up she said, was to go to the tea counter ten meters away.

Other passengers in the station persuaded her to go to the hospital, which she reached on foot!

Hundreds, probably thousands, had been in contact while she was contagious. At least a thousand were vaccinated in Karimganj and all 5,355 houses visited. Containment squads poured through the places she had traversed in her journeys.

Every train that had passed through the two stations where she stayed was tracked down. In all searches were conducted in 801 populated places, 31 railroad stations and 24,475 houses. The searchers found 48 cases of chickenpox but not one of smallpox.

On July 4, 1975, Saiban Bibi was released from the hospital and returned to her home in Bangladesh. She was the last human to suffer from the wrath of Sitala in India.

All through the course of the eradication campaign the reward for finding cases had been raised, reaching Rs1,000, a fortune in India, during Saiban Bibi's hospitalization. That amount was never paid. One year and six months after the extended campaign began, the eradication program had achieved its greatest triumph.

Grasset and Foege were relaxed one day in Clark's Hotel in Lucknow, looking out over the old town, then enshrouded in an early morning mist. Grasset said she thought that was the way children see India—beautiful, misty and still.

It was an India which—for the first time in its history—was free of smallpox. Sitala had been vanquished and India was beautiful, misty and still.

XV

Bangladesh

If it is true that every people should have its own nation then there is justification for the People's Republic of Bangladesh. That sometimes seems the only justification. No more ill-starred and wretched place exists on this world.

It has virtually no natural resources. There is natural gas offshore in the Bay of Bengal, but it has never been exploited. Its one exportable commodity is jute, a fiber in little demand and processed by Indian mills, beyond the control of the Bangladeshis.

To a large extent Bangladesh constitutes the massive river delta produced by the Ganges, Brahmaputra and the Meghna rivers. The silt carried by those three great rivers produces some of the richest, most fertile soil in the world, but that is almost useless to the Bangladeshis. Few parts of the country are more than a few meters above sea level and Bangladesh sits on "Cyclone Alley." Gigantic storms roil up the bay and then sweep across Bangladesh, producing devastating winds and floods that tear away the soil, the homes, the people. What Bangladesh does have in plenty is people, 83 million of them crammed into a land slightly smaller than Wisconsin, so that no matter how fertile the soil, there is rarely enough food, and famine is common.

"Bangladesh is the toughest place in the world," Brilliant says. Some demographers believe that by the year 2025 it will have more than 200 million people, if the present rate of population increase continues. The per capita income is $75. Infant mortality is 14 percent. Life expectancy is 35.8 years. "Fifty percent of all children die before reaching adulthood, and half from diarrhea.

"It's a grim, hard life. The hardening by fire, which the Indians call *tapasya*—a spiritual quest of life training—where you put somebody through the fires of hell so that they become hard and able to do anything . . . well, in Bangladesh it's involuntary."

It is no wonder that former U.S. Secretary of State Henry Kissinger called Bangladesh an "international basket case." In a country like Bangladesh smallpox is almost a petty annoyance. One storm—as history has proven—can kill more people in a year than any disease. Smallpox, in fact, was not the major problem it was elsewhere, particularly in India. Actually the disease had been eradicated in the former East Pakistan only to be reseeded as a fruit of the war whereby it gained independence as Bangladesh.

It is known that variolation was commonly practiced in the area. A newspaper in 1871 reported: "Outbreaks of smallpox are common. Inoculation is very largely practiced in the district, thus tending to spread the disease, and resulting in many fatal cases."

The inoculators were called *kabaraj*. By the end of the nineteenth century the British rulers tried to regulate them by insisting that they use government-supplied lymph for the variolation.

As in most places, various cures were devised. If no more effective than the others, at least they were more colorful than most:

> In smallpox three remedies are used—viz., (1) juice of the akan tree and mustard rubbed into the neck and throat; onion, plantains, and soaked rice are given as diet, with hot water fomentation and bathing; (2) a mixture of alligator flesh, honey and night soil; (3) juice of [two plants] are mixed together and poured down the throat, nose and ears.

The disease was known as *boshonto,* meaning spring in the Bengali language, because of its seasonal fluctuation.

In July, 1961, the Pakistani government climbed aboard the eradication bandwagon and ordered universal vaccination within a two-year period. Production was to switch to the freeze-dried vaccine and a program of follow-up control was to continue for at least ten years.

Severe setbacks immediately ensued.

Seventy-five million vaccinations were reported in the next three years and another sixty-eight million (including revaccinations) in the following three, but there was little noticeable decline in cases.

> Actually this programme has achieved the result of a control programme only [a government report stated]. The drawbacks suffered . . . were mainly of [an] administrative nature, set back by frequent cyclones and [an] influx of refugees from the neighbouring country [India].
>
> [Two years later:] In spite of all these high figures of

> vaccination and revaccination, smallpox continued to appear. The number of cases decreased to 71, an all-time low, at the end of the attack phase in 1964. In 1965, cases increased to 316, and in the year 1966, to 3,207. Obviously, the eradication campaign as it was conducted, was not so effective as we hoped.*

But in 1969, well before many other countries, East Pakistan switched to surveillance and containment. In East Pakistan the disease yielded to the new technique. In the first half of 1970 only four districts, Mymensingh, Sylhet, Bogra and Tangail, reported cases in any substantial number. After August, with but one case reported, the disease appeared to have been eradicated.

During the chaos of the ensuing independence struggle no smallpox was reported locally at all. But human intervention would reintroduce the terror, one of several almost Biblical afflictions to smite the poor Bangladeshis.

Islam came to East Bengal in the twelfth century with the Mongol invasion. Elsewhere conversion generally was resisted, but East Bengal embraced the new religion for the most part, while retaining its cultural identity.

In 1947 the Indian subcontinent became free of British rule. On Muslim insistence, two countries were created—India, where the population was predominantly Hindu, and Pakistan, largely Muslim. Pakistan in effect was two countries, separated by language and culture and by more than a thousand miles of India.

The cyclone of November, 1970, that brought the Brilliants to the subcontinent saw relief from the government centered in West Pakistan come in slowly and unsympathetically. The chief Bengali political party, the Awami League,

---

* As absolute numbers these figures bore no resemblance to reality.

led by the charismatic Sheikh Mujibur Rahman, gained a majority of seats in the Pakistani national assembly. The Awami were insisting on self-government for the eastern province and more control over its economy. The opponents seemed doomed to violent confrontation as there was no political party capable of bridging the gap between them.

In March, 1971, the Pakistani president, General Agha Yahya Khan, announced that he was going to delay opening the assembly, and the Awami were cheated of their victory. Rioting broke out in East Pakistan, general strikes were called in the cities and many non-Bengalis were murdered. On March 25 Yahya Khan ordered the army to put down the civil disorder, but the East Bengal regiment mutineed and destroyed bridges and railroads. The next day Rahman declared East Pakistan the independent nation of Bangladesh and was promptly jailed for treason. West Pakistani troops, mostly ethnic Punjabis, began a bloody occupation.

Hundreds of thousands were killed. The sides vied in matching atrocities, with the West Pakistanis apparently the winners. Women raped by the occupying troops were ejected from their homes by their husbands or fathers and left to starvation or suicide. Millions fled across the border into India as the terror reached unspeakable depths.

The Indians, who have trouble feeding their own people, put the Bengalis in large refugee camps near the border. The largest, with 200,000-300,000 people, was Salt Lake, near Calcutta.

One day an American going through the camp thought that he saw a case of smallpox. With such crowding under unhygienic conditions epidemics are a constant threat. The American notified D. A. Henderson in Geneva, who contacted Grasset in New Delhi. The Indian government assured her that it was only chickenpox, that there was no smallpox in Salt Lake, "not to worry."

Meanwhile the political situation between India and

Pakistan reached a nadir. On December 3 Pakistani fighter planes attacked Indian air bases in western India. The Indian government retaliated by recognizing independent Bangladesh and invaded the territory. A short, bitter war ensued.

The refugees remained huddled in the camps, waiting for the fighting to end. Television camera teams from around the world visited the camps to broadcast their plight. One crew went into a hospital run by Indian Caritas and photographed the patients. In Atlanta Foege watching the film on his home television set gasped as he saw the camera pass over someone who clearly was suffering from smallpox. Immediately he notified Henderson through CDC. On WHO's insistence a smallpox worker in West Bengal went to Salt Lake and found smallpox all over the camp.

Caritas ordered everyone vaccinated and instituted a fifteen-day quarantine. However, West Pakistani troops in Bangladesh surrendered on December 16 and the refugees poured out of the camps before anyone could stop them. They piled onto trucks and buses and later trains to head for their homes in their own independent nation. With them they carried the invisible fire of smallpox. The country that had just eradicated the disease was about to suffer its worst epidemic.

By March, 1972, eight districts were reporting smallpox. A search of a sixth of the villages in one *thana* (subdivision) showed 2,298 cases. Every case traceable had its origin either in the refugee camps or the transports back to Bangladesh. The scope of the problem quickly became clear. One out of every three Bangladeshis had fled for at least three months, roaming across the countryside trying to find food and shelter and safety from the soldiers. An estimated 9.8 million entered India. Now all of them were returning to their homes in a huge tumble of humanity, well beyond the control of what little government existed. Many of the homeless settled in *bastees,* ramshackle slums erected around the major cities. There they became sick, and being Muslims away from family who could help them, they decided to return home by any means—foot, train or boat. They spread their infection with them. The situation was complete chaos.

WHO and the Bangladeshi government knew that unless the situation could be controlled by September, the next year would see a completely uncontrollable epidemic of historic proportions. Some four thousand vaccinators, trained to use the bifurcated needle and freeze-dried vaccine, were sent to isolate the outbreaks. But nature, which seems to take a perverse pleasure in handing catastrophe to the Ganges delta, intervened. Drought and famine struck in July and August, and those Bengalis who had made it back to their homes took to the roads again, this time fleeing starvation. Again many carried the disease.

In August smallpox had spread to the *bastee* at the old railway yard around Dacca. The local sanitary inspector did not bother to make his report until October, and no effective action was taken. From the *bastee* the disease spread to the city of two million and also appeared in the transit camps set up to handle the homeless. Some camps had half a million people. The government shifted its emphasis to surveillance and containment in the urban areas and the camps, but it was a losing battle.

"Only about 10 percent of the cases were being reported," says Stan Foster, who was sent in by WHO as a short-term consultant. Besides, health workers were revealing only half the cases they knew about. Since mass vaccination had been the strategy, when smallpox occurred and was reported, the workers were penalized because they hadn't done their vaccinations so they tended to keep quiet, Foster says.

Some of the resistance to reporting stemmed from the

days of British rule, when the houses of smallpox victims were burned and their bodies sometimes cremated, which, in Muslim countries, is sacrilege.

Another problem was the forceable removal of smallpox cases to isolation camps. "If a Muslim is sick he wants to die at home," Foster says, "and so you have another reason why the public wouldn't report to the health workers. The reports had not always led to effective action in the past, so many people felt that there was no good in reporting smallpox."

The smallpox program had an excellent hard core of workers left from before the war, WHO found. At first there were only five surveillance teams. They went from market to market, showing pictures of smallpox victims and asking for information. The market surveys were productive, locating 80 percent of the cases in the catchment areas, and soon the number of teams was increased to fifty-five. Even with this expansion, the country could be covered only once every two or three months, and entire outbreaks were missed.

"Some of these people worked twenty-five days a month in the field for the full four years of the program," says Foster, who would eventually head the Bangladeshi WHO team. "Really, they were some of the most honest, hard-working people I have ever met."

The disease hunters soon increased to include all twelve thousand health workers in the country. Rewards were instituted to encourage reporting, but WHO made the tactical mistake of granting one reward only for each case reported. Health workers, therefore, would tell no one about the reward so that they could claim it themselves.

"We found that after we introduced the reward that the percentage of the people who knew about the reward only went up about 30 percent," Foster says. "When we realized that we had made a ghastly mistake, we doubled the reward, paying both the first person who reported and the health worker." That system worked. Knowledge of the reward, crucial for success in surveillance, went up to 65 percent and finally to 92 percent.

"Surveillance in Bangladesh went through four phases," he says. "The first was suppression and poor organization and all the reasons for not reporting. Then we added the surveillance teams, which did a really good job but just didn't do it fast enough. Then we started the reward and that improved it still more, but it wasn't until we got the house-to-house search going that we were really able to increase the knowledge up to 92 percent."

The house-to-house searches, however, still were not altogether efficient, because the health workers in the villages served several different programs of the ministry of health at the same time: registering couples, completing health questionnaires, distributing vitamins and contraceptives. *Then* they hunted smallpox.

The country was still in near chaos. Rahman, who was considered a saint by the Bangladeshis during the struggle for independence, found running the country an entirely different matter from freeing it. The economy was virtually nonexistent and the government didn't reach the village level.

In the spring of 1974 the government acknowledged its helplessness and turned to WHO for help to organize the health program and the smallpox efforts.

"We were in big trouble," says Foster, "and that was because there were three chains of authority and responsibility: there was a health chain, there was a malaria chain and a civil surgeon chain, and all three of these were separate chains of responsibility. . . . One night there was a change in administration in the ministry and I just told them, 'It ain't gonna work, we've got to have a single line of authority.' So they gave me one night to come up with a structure and I totally redesigned the line of authority and responsibility so that at each level there was a single person responsible.

"A first step was a monthly reporting system, which initially listed cases and deaths but eventually went to the system where we started looking at outbreaks. . . ."

The searchers reported the number of cases, both active and closed, the interval between the first attack and detection, and the interval between detection and the last attack. Every month there was a meeting, at either the regional or the national level, which was dedicated to failure analysis: where the surveillance system broke down; how it happened. The WHO staff gained many insights from the local workers.

Slight modifications could make a sizable difference. The staff found, for instance, that seventeen separate outbreaks were caused by families from one village visiting victims in another. Yet when they looked at the list of "visitors" provided by the victims' families, they did not see the relatives' names. Family were not "visitors." The wording of the questionnaires had to be changed.

The new system was not perfect but it was making inroads. By August of 1974 smallpox was down to two subdivisions, Kurigram in Rangpur and Jamlpur in Mymensingh. Foster predicted the end of smallpox in Bangladesh by December. He had not taken into account the forces of nature.

In August and September the floods struck, the worst in twenty years. The hardest hit sections were, naturally, Rangpur and Mymensingh. Crops and homes were destroyed and, once again, the people fled.

They poured into the *bastees* around the cities, looking for shelter and work. By December smallpox had broken out in the Dacca *bastee.*

"I arrived in Dacca and visited the hospital there," Grasset says. "I rang up Foster and said, 'We've seen smallpox patients at the hospital here and they've been here for four days.' " Nothing had been reported from the hospital to the eradicators.

Foster, furious, demanded an explanation from the hospital officials. He was told that the telephone hadn't been working. The hospital was a three-minute walk from the smallpox office!

Smallpox was found in the city outside the hospital and the *bastee.*

Now it was time for human stupidity to multiply the problems. In January, 1975, just after the outbreak was discovered, the government decided to raze the *bastees* at Dacca, Khulna, Rangpur and Chittagong. Perhaps 100,000 people, including many with smallpox, were dispersed around the countryside. All hell broke loose again. This time the government really was shaken.

A report issued by the government and WHO frankly stated:

> The disruption of the progress toward smallpox eradication at a time when neighbouring countries were nearing this goal, questioned the credibility of the whole concept of eradication. Further, it questioned the credibility of the Bangladesh government, as an effective implementor of health programmes, and of WHO as a technical assistance organization. In February, 1975, a presidential directive was issued which declared smallpox a national emergency and ordered the mobilization of all available resources to assist in the smallpox eradication program. Bilateral and international aid was also requested.

The situation was dire. Grasset, who thought she had seen everything during the war in Biafra, was appalled at what she witnessed in Bangladesh.

"I was with D. A. [Henderson] going to the isolation ward for smallpox. At the time when smallpox was rampant in Bangladesh they had a drought in the north. To see children starve and at the same time be full of smallpox from top to bottom, it's just the most terrible, ghastly sight I've ever seen.

"In the hospital a child all of a sudden began wailing and just fell back and died. What horrified me was that the mother was wailing—you know how those women can wail—but not touching her child. The child was such a mess of pus that even its mother didn't take it in her arms as it died!"

That night Grasset and Henderson went to the home of an American WHO administrator on loan from CDC. "He had a child, about ten months old," Grasset says. "It was a lovely, fat, white, good-looking baby." The home was half a kilometer from where the other child died. The contrast shook her for days.

With the program in India winding down with victory certain, WHO began massive shifts of personnel and money to Bangladesh. Henderson, Grasset and Foege in Dacca plotted the strategy. More than the mere fifteen international epidemiologists present in January, 1975, was the prime necessity. With a thousand outbreaks anticipated in the spring and the number of outbreaks a single epidemiologist could possibly handle twenty, speed was essential.

The first twenty-eight arrived in Dacca in February, all on the same plane.

"Obviously we went from a medium-size program to a bigger program overnight," Andy Agle, an administration officer, said. "We were not in a position really to accommodate this many epidemiologists. We just did not have the administrative capacity, the logistical capacity, we didn't have enough vehicles. In fact, just to meet the plane carrying these twenty-eight epidemiologists we had to rent a city bus. We rented the bus for a whole week while they were training in Dacca."

More reinforcements followed. Every other Tuesday a Thai Airlines flight from New Delhi would bring in another half dozen internationals. They would be trained until Saturday and promptly dispatched into the field.

Vehicles also arrived—new Mahendra Jeeps from India supplied by SIDA, the Swedish aid agency.

"When I arrived in September, seventy-four, we used to get our check from New Delhi to take care of local costs, and I got a check for $20,000 each *month.* In the spring of seventy-five, when we were going great guns, I was getting a check every *week* for $75,000," Agle says. "The administrative staff also increased. I guess I kind of insisted on that. We went from having just an administrative officer, to having an administrative officer with a transport officer, a supply officer, two expatriate mechanics, a communications officer, who handled the radio, and a finance officer."

This was to be the last general call to arms for the army fighting together under the battle flag of smallpox eradication. Everyone sooner or later showed up. Zikmund and Selivanov, two of the four original epidemiologists in India, arrived. Larry Altman came over to cover the campaign for the *New York Times.* Basu, Brilliant and Foege popped over from New Delhi. Henderson's second-in-command and heir apparent, Isao Arita, and Henderson's first boss at CDC, Alex Langmuir, were among the arrivals. A number of scientists who had devoted part of their lives to the study of variola, perhaps sensing the end of the species, also dropped by to watch, men like Britain's Keith R. Dumbell of St.

Mary's Hospital. Geneva sent in Alan Schnur and John Wickett and other veteran administrators and vaccinators.* All that was learned in earlier campaigns was now applied to Bangladesh. It became the best-turned, best-run eradication program ever.

The epidemic raged despite the army fighting it. By the middle of March, 3,708 cases, 67 percent of the world total, were in Bangladesh. There were 991 infected villages.

Surveillance and containment was tightened up. The object was to find at least 80 percent of the cases within fifteen days of an attack. Containment was to be 100 percent within twenty-one days of detection. WHO demanded that at least 90 percent of all cases be traced to the source.

A health worker was posted to each infected village. Six inhabitants were hired to help and, as in India, guards were posted at the infected house. Tally forms were kept to make sure that everyone within a half-mile radius was vaccinated. An epidemiologist was assigned to each fifteen villages to enforce the procedures.

Unlike Hindu India, there was no smallpox cult in Muslim Bangladesh, no goddess to contend with. But, as a disadvantage, ritual isolation of the victim was not practiced. Indeed, Bangladeshi families seemed to feel an obligation to visit their afflicted relatives, usually after dusk, and run the risk of exposure.

WHO took over a fourteen-station radio network from the U.N. Disaster Relief Organization, which was leaving Bangladesh now that its wartime task was done. WHO's staff could keep in contact with every district. Every morning Dacca went on the air to check with its district people, and

---

* Secretaries in Geneva like Britishers Sue Woolnough and Celia Sands were now putting in sixteen-hour days just keeping lines of communication open.

in this way they kept track of the virus.

The epidemiologists scoured the rice paddies and marshes of the delta, cruised the waterways by boat, rode on the top of trains jammed with people, crossed rivers on the fragile bamboo bridges, or flew above the palms on Russian helicopters. (At least one epidemiologist was killed in the effort, Dr. A. B. M. Kamrul Huda of Chittagong, who drowned when his speedboat overturned while he was investigating an outbreak on Sandip Island on March 3.)

Searchers routinely examined the bodies in cemeteries.

WHO still worried that some Bangladeshis were concealing outbreaks. A reward equivalent to $20, one month's pay for a prosperous Bangladeshi, was offered as a lure.

The Bangladesh effort was handicapped by a lack of human resources. "India had a tremendous storehouse of national talent," Brilliant says. "There were Indian epidemiologists, who, as Dr. Sharma would say, were 'retired but not tired' that we could call back. In Bangladesh that kind of manpower wasn't available and their program was almost as big as the one in India in terms of the number of foreigners."

Nevertheless the measures seemed to be working. The 939 outbreaks in May, which was supposed to be the high period, were less than the number reported five weeks earlier. WHO believed that if it could get through the monsoon season with less than a hundred outbreaks, the disease could be stopped in the autumn.

For once nature was kind. The weather held and the crops flourished. There was enough food so that the population did not disperse. There were no floods, no calamities.

By August 2 active cases in the country totaled only 47. With India now smallpox-free, the last cases in Asia, and the last of variola major, were all in Bangladesh. But now politics intervened.

On August 11 Rahman was assassinated in a political coup.

The first to react were the Indians, who feared that disorders would drive infected Bangladeshis across the border.

As it is impossible to seal off the largely unfenced, unguarded, and unpatrolled Bangladesh-India frontier, the Indians decided to set up a *cordon sanitaire* ten miles on their side of the line. Every village within the zone was searched repeatedly. The Indians generally knew where Bangladeshi refugees would go; they would seek out friends and relatives who were also Bengalis, and the places where these people lived were also searched thoroughly. People who were found to have crossed the border were checked for vaccination scars and asked where in India they were going. The health officer at the destination was advised that they were coming and told to keep an eye on them.

Anyone arriving at an airport from Bangladesh was checked for vaccination. Health officials refused to accept the yellow WHO card because it could be faked so easily and ordered everyone, including VIPs, to take off their coats and show their scars. Some Dacca bureaucrats and officials thought that they were too important to be subject to such indignities.

"Our main objective was to make the Bangladesh government conscious of its responsibilities," Basu said.

There were no importations. On August 15, Indian Independence Day, India celebrated its smallpox-free status with a ceremony. Smallpox stayed across the border.

By September 13, the end of the monsoon season, there were only seventeen known cases anywhere in Bangladesh. Victory seemed assured.

"We are cautious because, in the past, when we had no smallpox cases in our country, smallpox occurred again," the health minister, Abdul Mannan told Larry Altman. "We use statistics provided by WHO as our official statistics, and we hope we can bid good-bye to this dreadful disease later this year."

The disease would not go so easily. It had one wry joke left to play. On September 15 a two-year-old girl, Sokina Begum in Shekhpara village in Chittagong, developed a smallpox rash. Her village was sealed off and everyone was vaccinated.

Nothing more happened. There were no new cases. WHO continued to stalk the countryside, but still there were no cases.

On November 14, 1975, WHO declared that smallpox had been eradicated in Bangladesh. The WHO team joyfully scheduled a victory party.

"I was always very wary of having parties to celebrate the end of smallpox," says Agle, remembering the Nigeria fiasco. "But we did just that, we had our party." It was held in the WHO guest house, whimsically named WHOSE house (for World Health Organization Smallpox Eradication). "Quite a few people got pretty well juiced."

The next morning the slightly hung-over headquarters team went to the office and sifted through the congratulatory telegrams that had begun to come in from the world.

CONGRATULATIONS FOR GREATEST ACHIEVEMENT, the first one read.

CONGRATULATIONS ALL DELIGHTED, said the second.

ONE ACTIVE SMALLPOX CASE DETECTED VILLAGE KURALIA UC SOUTH DINGALDI PS BHOLA DATE OF DETECTION 14/11/75 DATE ATTACK 30/10/75 CONTAINMENT DETAILS FOLLOW, said the third.

They were not done yet.

The Bhola island case had been discovered by Bangladeshi health officers and had been suppressed. WHO felt not a little foolish.

Smallpox workers in Bangladesh have discovered new

> cases of the viral disease, forcing the World Health Organization and the Bangladesh Government to retract a November 13 [*sic*] announcement that smallpox had been wiped out in Asia [wrote Altman for the *New York Times*].
>
> Disappointed World Health Organization officials have backed down from their earlier announcement that Asia was free of smallpox for the first time in history [said the *Wall Street Journal*].

WHO had to find out what had happened and how serious the outbreak was. Bhola was one of the most remote islands in Bangladesh but if the disease could be festering there it could on other islands as well. A WHO team was assembled quickly to make the difficult trip.

Brilliant flew from New Delhi to Dacca and continued to Barisal by a paddleboat so crammed he had trouble moving about the deck. In Barisal Brilliant went to the area's civil surgeon, who had been an obstruction to the program for years. He had sternly resisted reporting cases in order to keep Dacca and WHO out of his territory. Now he wanted to make peace and offered Brilliant meat from a calf he had had slaughtered as an offering for the Islamic holiday of *Eid ul azha.* It was the first meat Brilliant had eaten in four years. He loved it. Peace was achieved. The surgeon and Brilliant began to search the island district for more cases.

Meanwhile the WHO team drafted from Dacca had arrived at Bhola. They had nurtured the hope that the case was only chickenpox. The first look at the three-year-old girl victim, Rahima Banu, ended the fantasy: "It was a clear case of smallpox. The scar distribution was typical; some smallpox scabs could be seen on the foot sticking out from under the blanket."

A lab later confirmed the diagnosis. Meanwhile the entire island had to be sealed off to prevent the virus from escaping. Like most Asians, Rahima was suffering from variola major, and if that was the last outbreak then the only remaining variola major left free in the world was on Bhola.

The island, one of the last of the delta before it spills into the Bay of Bengal, has an area of 2,600 square kilometers and a population of about a million. It was connected to Barisal and the rest of Bangladesh by a ferry service. At last Brilliant had reached the goal of himself and the Hog Farm after the 1970 cyclone!

Bhola is in the groove in Cyclone Alley, and averages four storms a year. When the cyclone of 1970 struck, the only survivors in some sections were the people young enough and strong enough to climb the tallest trees and cling up there for thirty hours. Many of them suffered from abrasions on the chest from clasping the rough tree trunks—doctors learned to call these the "clinging syndrome." Below them twenty-foot tidal waves rolled in tiers across the land, sweeping away whole villages.

Many of those clinging to the trees ran out of strength and lost their grip, falling into the maelstrom below.

Even in normal times getting around Bhola island can be difficult. The main transportation link in many areas is the bamboo bridge, a single length of bamboo stretched from the embankment of one rice paddy to another. They are extremely perilous, and more than one WHO worker fell off a bridge while trying to approach a village. Few epidemiologists have the agility of high-wire experts. The locals would stand around and watch every time a foreigner tried to cross one of their bridges, an act that was almost a silent test of courage.

But the Bangladeshi team was able to trace the final outbreak. Between August 1 and the premature celebration of November 14, there had been 141 cases on Bhola, including 33 deaths. The outbreaks generally were over by the time that the local health workers found them, although two were still active. Some outbreaks had not been reported to the central government.

On November 6 an epidemiologist and his team posted to Bhola the month before verified an outbreak of twelve cases in the village of Kathali and ordered containment within a five-mile radius. While drinking tea in a market he heard rumors of another outbreak, in West Joynagar, three miles to the south. A health worker earlier had concluded that it was measles. The surveillance team reinvestigated and found unmistakable smallpox, including three deaths. The Kathali outbreak could be traced to an importation from Chittagong, but there was no clue concerning West Joynagar.

A little girl, Bilkisunnessa, told the surveillance team of the outbreak in nearby Kuralia. There they found Rahima Banu, who apparently had caught the disease in West Joynagar. The team notified Dacca, spoiling the victory party.

The massive staff effort rushed personnel, motorcycles, bicycles, speedboat engines, drums of fuel, kerosene lanterns, loudspeakers and vaccine kits to Bhola. Everyone within a mile and a half of Rahima's house, all 18,150, was vaccinated. Vaccinators worked night and day, yielding to no excuses or resistance. Rahima was isolated in her house and her family was paid to stay inside with her. Guards were posted to keep everyone else out. It was possible that her little body contained the sole survivors of the virus variola major in a wild state, and the eradicators wanted them to go no further.

The five-mile radius around her house was searched repeatedly by different teams, each assessing the work of its predecessor. Seven markets and nine schools were visited. Every possible contact was checked and double-checked to see whom they had come in contact with in turn after their first exposure.

Transportation was still the greatest problem. Many villages near the shore were accessible only by foot or bicycle.

For others that could be reached only by boat the program brought in a motor launch. A second vessel, privately owned, was also hired. It made one trip, but the crew was so frightened by the surging, bloated, postmonsoon river, that it returned to Dacca the next day.

Checkpoints were set up at bus stations, ferry landings and crossroads. Everyone was looking for rash-with-fever cases which could turn out to be smallpox. Posters, handbills and pamphlets advertising the $20 reward were printed. Loudspeakers and radio stations broadcast the plea for information.

To counter the rumors that anyone found with smallpox would be burned with his house, after Rahima's family ceased to represent a threat of infection they were drafted to go to the villages and dispel the fear. Rahima's father, the illiterate, landless laborer Waziuddin, was the most persuasive.

No more smallpox was found on Bhola or the mainland although the searchers did turn up 398 cases of chickenpox and 326 of measles.

On Christmas Eve Larry Brilliant, who had failed to find any cases away from Bhola, encountered the uncontrollably crying Rahima. She was distressed over all the foreigners who were coming to prod her and stare at her still-scarred body. She clutched her mother.

No one knows if her family understands what Rahima represents. Eventually they were invited to another party in Dacca and hosted at a luncheon. Clearly they had never been to anything quite that civilized in their lives, and the little girl was not pleased with the attention.

But this party went uninterrupted. No new telegrams arrived to dash their hopes. Rahima Banu was the last human to be infected with wild variola major virus, the last person in Asia to suffer from a natural smallpox infection. Ever.

The army put together by the World Health Organization was slowly dispersed. Only one more battle needed to be fought, against variola minor in Ethiopia and Somalia, and that did not require great numbers in anyone's judgment.

Smallpox had been defeated and only cleanup operations were left. They would prove to be more difficult than anyone had imagined. The disease did not go easily.

XVI
The Long Last Stand

It was not unreasonable that many in the smallpox army thought that the last stand of the virus in Ethiopia would be anticlimactic. The Indian subcontinent was the biggest, most complex and most difficult of all the campaigns. There they were fighting the deadly goddess. Ethiopia, on the other hand, was reporting only a few cases of variola minor. With surveillance-and-containment techniques refined, and with smallpox-eradication programs down to rote, it appeared that Ethiopia would be cleared of the disease quickly and the world would know it no more.

It was tougher, took longer and was far more dangerous than WHO ever had anticipated.

The eradication program in Ethiopia had begun in 1971, two years after most other countries. It was at best a shadow program. The Ethiopian health ministry was engaged already in a losing battle to eradicate malaria and it was very reluctant to divert any resources to another disease, especially since smallpox in the country was a relatively minor problem.

Under pressure from WHO, which emphasized that the disease could not be eradicated if there was still a nest of viruses anywhere in the world, the Ethiopian government agreed to a small program. It would pay the salaries of twenty health officers and sanitarians—if they could be found—but everything else would have to come from WHO.

WHO was committed elsewhere and had little to give. The only outside help WHO could get was a pledge by the United States to supply seventeen Peace Corps volunteers. WHO also managed to provide twenty-one vehicles and $120,000. Some auxiliary personnel also was available.

Considering the resources, the nature of the country, the geography and the population, the task seemed impossible.

Ethiopia is large and quite empty. It is about twice the size of Texas and has more than twice the population, twenty-four million people. The part of the country to the south and east can be described kindly as desolate. The Great Rift Valley, where two huge continental tectonic plates were born, cuts through the region and plunges below the Red Sea just where Ethiopia meets the Somali Democratic Republic at the tiny country of Djibouti. The terrain is so harsh, so primitive that geologists go there to study the creation.

The southeast is the Ogaden Plateau, rugged and empty. Nomads, perhaps the most independent people on earth, silently cross the desert following water for their flocks, pitching their huts for a week or two and then quietly moving on. Leopards and lions roam through the night. Malarial mosquitoes break the silence. Vultures sit on the sparse tree limbs waiting for something to die.

More than half the people of Ethiopia live more than one day's walk from the nearest road. Outside of Addis Ababa and the provincial capitals there are no telegraphic or telephone communications.

Well-watered highlands constituting much of central and western Ethiopia and some other areas are home to about 70 percent of the Ethiopian population. Deep gulleys and ravines are frequent. Small groups of houses cluster together on the edges. Such places gave the eradicators the most difficulty, not only because of the rugged terrain but because many of the people absolutely refused to be vaccinated. They practiced variolation and would not be convinced that there was anything better. They were quite capable of violence to have their way, as the eradicators would soon learn. And if the locals were not difficult enough, the countryside was plagued with *shiftas,* roving armed highwaymen like the *dacoits* of India.

Because the disease was relatively mild, and because the obstructions to any vaccination program were so great, very little work had been done before 1971 and it was estimated that perhaps only 5 percent of the population was protected. With less than five hundred cases reported in any one year, immunization seemed relatively unimportant. (When a reliable reporting system was established the true figures would be one of the more unpleasant surprises.)

With resources severely limited WHO and the government set up a modest plan that spread the workers around as much as possible and attempted to give some protection for Ethiopia's neighbors. Two teams were sent to each of the fourteen provinces in the country. Each consisted of two or three people and one vehicle. The job of the teams was to try to convince the health services in the provinces to report smallpox and themselves to respond as necessary. In those many areas where population clusters were nearly inaccessible, they went along the roads vaccinating anyone they could find. Getting beyond the roads would have to wait for another day.

The most energetic efforts were in the provinces in the west and north bordering Sudan and those in the south bordering Kenya. WHO hoped to continue to keep both those countries free of the disease. The Ogaden Plateau which bordered on the Somali Republic generally was left alone because the resources to search the desert and chase the nomads simply weren't available.

The vaccinators promptly discovered that the disease was much more common than had been believed. One officer visiting a school with pictures of smallpox victims to ask if any of the students had seen a case found more work than he could handle. Some of the outbreaks were so widespread that the smallpox teams could vaccinate only around the edges and gave up any attempt to try to vaccinate within the area itself.

The mildness of the disease proved to be a problem, not only because so many people refused to take it or the eradication program seriously, but also because sufferers would not be incapacitated and could move about spreading the virus. One infected man, leading a herd, walked 300 kilometers and crossed into Kenya. The fatality rate was 2.1 percent, even lower than elsewhere in Africa.

The first year the smallpox teams found 26,329 cases, fifty times more than the government had ever reported but probably less than 10 percent of the actual number. Vaccinations were administered to 3,385,600 people. Smallpox was rife in Ethiopia and eradication would not be easy at all.

The next year, 1972, WHO got more help. The U.S. committed more Peace Corpsmen, Austria sent four persons from its overseas volunteer service and Japan sent fourteen persons and several vehicles.

The total resources increased to seventy-five trained persons and fifty vehicles. As a sign of definite progress, control was gained of a major chain of infection in Eritrea, the relatively developed province along the Red Sea coast.

The typical cyclical character of the disease was like that in Asia. After the rains in September, cases increased in number, and smallpox literally exploded. One team of Peace Corps volunteers working in Gemu-Gofa Province in the southwest reported finding smallpox in practically every village. Some chains of transmission could be blocked, but often only mass vaccination could cope with the disease.

And, as always, transportation was a horrendous problem:

> Only one all-weather road exists in the province. The other roads are not only hazardous to drive, but repeatedly induce mechanical failure of the vehicle. Driving skill and knowledge of the trails has, of course, improved over the past year, but the major reason for fewer repairs during last year is that the vehicle is now used only for ferrying people and supplies to some focal point easily accessible by car. Almost all surveillance is now being done by foot or mule. Planes and boats are sometimes used in order to get team members to especially remote areas, where they walk or ride to the outbreak. The increased use of mules and walking has increased the thoroughness of investigation and surveillance.

Whenever the teams struck out by foot they had to leave many amenities behind, amenities which sometimes included sufficient food. Some regional officials provided letters ordering their local subordinates to supply housing and food. In a place where survival is marginal at best, this undoubtedly proved to be a burden to the villagers, one that they did not always accept graciously. But the letters lent great authority, and many local leaders put up the smallpox teams in their own homes and ordered the villagers to line up for vaccination. Sometimes the headmen would go door to door with the team to make sure that everyone knew that these strangers had the force of government behind them.

But many villagers still refused vaccination, some violently. In one WHO document there is the cryptic report that "many workers were suffering at one time from human bites."

Some techniques were better than others:

> We find, for example, that bright, colourful, preferably grotesque pictures of smallpox at least make the people curious; and while they may not want vaccination initially, they will often crowd around to see the picture. In this manner, some always get vaccinated and, hopefully, these can be encouraged to tell the rest that the vaccination does not hurt. In general, having a crowd around saves countless repetitions of the same assurances and explanations, encourages others to come over and see the cause of the gathering, and reinforces whatever group pressure there is to accept vaccination.

"Police," the eradicators found, were more of a hindrance than a help, "often frightening many and transforming 'getting a vaccination' into a game of hide-and-seek for the rest."

Attempts to initiate a reporting system did not succeed very well. The Gemu-Gofa smallpox team * was reconciled to the fact that they must "personally walk around the province" if they wanted to be sure they were beating the disease.

Attempts to hire locals, the so-called "army of young vaccinators," were abandoned because the WHO workers

---

* The Gemu-Gofa team consisted of two Ethiopians, Ato Girma Tilahun and Ato Kassa Mondaw, and two Americans, Daniel Kraushaar and Scott Holmberg.

lacked the expertise and experience to supervise large numbers of untrained people under difficult circumstances.

One of the Peace Corpsmen in the program was Alan Schnur, who would show up later as a free lance in the Indian campaign. Schnur learned the smallpox business in Ethiopia while also learning how to get along in one of the most godforsaken places in the world. (However, he says, "I find some of the most godforsaken places a lot nicer than parts of New York City.")

Schnur learned to work alone, carrying a sleeping bag, a change of clothes ("in case I fell into a river") and sometimes a book. Ethiopia is also where he learned the value of keeping a roll of toilet paper handy.

"That's mandatory," he says. "Once I was down to the bitter end; we were coming out to replenish our vaccine supply. I had diarrhea and a cold, and just as I got to the last piece we happened to reach the town, where I could buy some more.

"I did use a couple of smallpox forms once. We always used form Number 1 because we had so many."

Schnur sometimes would carry canned food like sardines, "in case I got stuck out somewhere. Most people carried a tent and cooking equipment, but I was too lazy. I preferred to sleep in somebody's house and eat his food. I think it worked out well."

Schnur says that the native food was quite safe because it was thoroughly cooked. Any contamination came between cooking and serving. He survived all of his meals.

One trick the vaccinators learned in Ethiopia (as elsewhere) was to get in and out of a village before the reactions to the vaccine erupted (about three days). Many people tried to lie their way out, claiming that they had been vaccinated already or that vaccination would make it impossible for them to work in their fields.

In the highlands resistance was heightened by the persistent rumor that the vaccinators were drawing blood to use in airplanes. The locals believed that jets ran on human blood.

Getting from one place to another in the highlands was the major chore. Sometimes it took eight hours for Schnur to go from one town to another, a trip that was later completed by helicopter in ten minutes.

Schnur dressed like a native, in the long skirt of the nomad. At night he would roll it up over his shoulders and use it as a blanket. During the day it offered protection from the sun.

"I changed clothes every two weeks, whether they were dirty or not. I was too lazy to wash them often so I just wore one set of clothes for two weeks and changed into the clean clothes for the next two weeks, and after a month I took them to somebody to wash.

"People didn't look at your clothes too closely. The water in Ethiopia was cold and muddy."

Part of Schnur's Peace Corps training was in Amharic, the most common language in Ethiopia. He would spend nights sitting up with his hosts talking about life and answering questions, frequently with considerable exaggeration, about America.

He carried some money, but usually in coins, because the natives did not trust the government's paper money. He would take the Ethiopian equivalent of $100 in quarters. A number of vaccinators were robbed by the *shiftas.*

For entertainment he had a little portable radio. The local people loved it, gathering around to listen, many for the first time in their lives.

But on foot or mule, Schnur and the other young eradicators brought the disease under control in several provinces. The number of reported cases in 1972 decreased to 16,000, with slightly improved surveillance. About 20 percent of the cases were not being reported by official government sources. Some 3.2 million Ethiopians were vaccinated.

In 1973 cases went down 68 percent even though the

WHO staff was still fewer than a hundred. Addis Ababa had only fifty-three cases, all importations. The only setback was when a local drought and famine touched off a mass migration that spread the disease to two other provinces, where it was quickly controlled.

In 1974 Ethiopia received additional help from some of its neighbors. Sudan sent fourteen men and three vehicles to the province of Gojjam. The French furnished forty-five men, twenty vehicles and three helicopters from the then Territory of the Afars and the Issas (now Djibouti). Kenya dispatched crews across the border. The campaign showed some progress as the number of outbreaks after the September rains was down drastically. Now, however, human affairs would intervene.

The reign of the venerable Lion of Judah, the Emperor Haile Selassie, once a symbol of courage to the world, had been disintegrating for years. His regime was corrupt and unresponsive. His time seemed near anyway, but the *coup de grâce* was the government's attempts to cover up the 1973 famine and its inability to provide aid. In March an army mutiny forced the emperor's cabinet to resign and the military took effective control of the country. Selassie's efforts to mollify the army and to retain any power, even nominally, failed, and in September the Lion of Judah was deposed. In November the military junta confirmed its power with a mass execution of sixty prominent Ethiopians. Almost every night the sleep of the living in the capital was interrupted by gunfire. Almost every morning the bodies of the dead littered the streets.

In this kind of chaos government is impossible and running a health program becomes enormously difficult. And this was not the only trouble facing the country. The province of Eritrea, once a separate colony of Italy, had been in revolt since 1971, and the civil war was escalating considerably in 1974, further diverting attention and resources.

In the south the centuries-old dispute between the Ethiopians and the ethnic Somalis in the Ogaden Plateau grew worse. The Somali government was supporting the rebels more openly than before. The Somalis claimed that the Ethiopians were waging a war of extermination against the nomads in the desert. What had been a hit-and-run rebellion showed every sign of turning into open warfare.

All this increased the level of violence in an already-violent situation. At least two Ethiopian health workers were shot and killed when they got too close to the action, and WHO pulled everyone back whenever the battle started. Since some of the areas of the greatest violence were also endemic regions, the withdrawal slowed the program considerably.

But despite these obstacles, the new vehicles and helicopters permitted the staff to get to areas it had been forced to ignore in the past, and by the end of 1974 the number of cases had gone down 20 percent, to 4,439.

Ethiopia until then had been a relative backwater in WHO's smallpox efforts compared to the massive campaigns in Asia, but with almost all of the subcontinent smallpox-free by early 1975, WHO began to turn more of its attentions to the country. The whole complexion of the campaign changed. The U.S. officially withdrew all of its Peace Corpsmen, as did the Austrian government its volunteers. Western support for the smallpox program, however, did not end. The U.S. sent two helicopters in addition to its two previously in the country. More vehicles and campaign equipment were also sent in from abroad.

Highly trained WHO epidemiologists, shifted from India, began to appear in Ethiopia. They started a program to recruit Ethiopian students as searchers and vaccinators. The government provided more of its health workers. The WHO budget was nearly tripled, from $850,000 to $2,243,000. They also found the Ethiopian director, Ato Yemani, was first rate.

With the easily accessible places now apparently smallpox-free, the program turned its attention to the highlands, the only known endemic area left. It would be the hardest, most violent, most difficult area ever encountered.

The natives were serious about resisting vaccination, as one WHO pilot, Robert Francis Lavack, found in January, 1975. He was flying a Hughes 500C chopper from Dabi, Gojjam Province, picking up vaccinators. He had put down at Endebago Zion in the mountains about 3 p.m. to pick up some more passengers. He was revving up the engine when the aircraft was attacked.

A hand grenade was thrown under the helicopter from about fifteen meters. It exploded, puncturing the fuel tank. The aircraft was enveloped in a sheet of flame. As the people inside began to leap out, rifle fire opened on the craft.

An Ethiopian guide, Shiferaw Gelaye, pulled a 7.65 mm. pistol and fired back. That little resistance was enough to end the conflict. Lavack says that he saw only one person fleeing, but Gelaye claimed that he was firing at three.

The WHO team managed to get back to Dabi safely but only with the help of a half dozen armed men.

Life was not uneventful in other ways. One letter sent from Geneva to the chief WHO epidemiologist, K. L. Weithaler, cryptically reports:

> We have received your cable dated 2 October 1975 pertaining to the damage to the office building resulting from the explosion which took place recently. We are happy to hear that nobody was injured and that in spite of these difficulties the work for the programme is continuing. . . . Needless to say that your work is appreciated in these circumstances.

Getting equipment in and out proved to be one of Weithaler's chief concerns:

> Things are getting a little bit tight these days [he wrote Geneva in February] because the helicopter which was supposed to arrive on 15th February has not yet arrived; and this means that there has been thirteen days delay up to now. According to the latest information it seems that this lousy apparatus is stuck in Nairobi. Because of this, poor Ciro * has to work Merha Bete by mules, while our helicopter in the neighbouring Borena does, of course, its utmost to support those teams. At any rate the latest figures show clearly that we are really working very hard in the last endemic areas of the country. As time is passing very fast we are in doubt whether we would win the battle by the coming rainy season or not. The continuous delay of the helicopter operation caused a very serious hindrance for the project and it is evident that even the most dedicated work could not make up for the lost time. . . .

Local conflicts, such as those in Eritrea and the Ogaden, also hampered the work, and Weithaler did not eradicate the disease by the 1975 rainy season.

The shifting of experienced WHO workers from Asia steadily continued, and the local staff was increased to 1,200. The U.S. sent more money, a gift of $3 million. Surveillance and containment began in earnest, despite resistance from the local people and the continuing problems of safety. Helicopters saved the day, but not without some risk.

In May a helicopter carrying Weithaler and three others ran out of fuel in the air over Bale Province after it became lost above storm clouds. It made a safe landing in the bush:

> Due to some lucky chances like sufficient food, water . . . camping equipment as well as the long lasting Af-

* The "Ciro" referred to here is the same Ciro de Quadros that Henderson so admired in Brazil, now working for Geneva.

rican experience of the pilot [identified only as a Mr. Tanner] and passengers, the incident did not become a disaster. There was enough water to drink and food to eat, and a minimum requirement for sheltering the crew [*sic*]. The helicopter was equipped with a special radio signal-giving emergency unit which in the end made relief actions successful.

But not easily. The other radios in the chopper could not pick up any station in the country and the first contact was with an Ethiopian Air Force plane flying overhead on the second day after the crash-landing. The Ethiopians homed in on the emergency signal.

A WHO helicopter, piloted by Bob Lavack, brought fuel to the downed aircraft so that it could resume its flight:

> Finally, it should be mentioned that no one of our passengers got injured or experienced any health troubles except having been bitten by hundreds of mosquitoes and having been surrounded by hungry hyenas. The morale of every passenger was excellent and everybody tried to face the facts as they were.

There were also several cases of smallpox workers being kidnapped by *shiftas*. Some were held hostage for a number of weeks.

There were other minidisasters. Serious flooding washed out eradication efforts in the Wabi-Shebale flood plain. *Shiftas* terrorized smallpox workers in Bale, causing work to be suspended for several weeks.

Yet the disease seemed to relent. On August 9, 1976, a three-year-old girl, Anina Salat, in the nomadic village of Dimo in Bale, erupted in a rash. Despite extensive searches in the area and in the accessible regions of the Ogaden, no more cases were found. For one brief golden moment WHO thought that it had achieved its final victory and that the little girl was The Last Case.

She was not. The virus had not yet made its last stand. It was hiding in the desert among the beautiful Somalis of the bush.

The nomads of the bush, like most Somalis, are very tall, slim and graceful with strong smooth features, high cheekbones, high broad foreheads and easy smiles. Their skin is coffee-colored. They move with a dignity and flow that bespeaks pride and independence. They may be among the most beautiful people in the world, both men and women.

They travel in clans and family groups, following the water. The men and boys drive the sheep and goats ahead of them, the women with the camels and the caged chickens follow.

When they find a water hole and a place to settle for a time, the women put up the small round houses of wood and thatch. The round peak is topped with a beer or soda bottle.

There is no furniture; there is no room for any. The adults sleep on thatched platforms on the side away from the open door. The children sleep on the floor. All cooking is done outside, even during the rains.

Each family group has two huts. One is for the people, the other for their few possessions and the chickens.

Each cluster of huts is surrounded by a *haro* of thornbushes. The *haro* is circular, with one gap to admit people. At night the gap is plugged with a thornbush door. The *haro* keeps the family's animals in, the leopards, lions and hyenas out. It is not really designed to prevent human attacks, which almost never happen in this desert. There are no locks on any doors.

There are few possessions to steal. By Western standards these are among the poorest people in the world. By more sensible standards they are among the richest. No one with forty or fifty sheep, cattle or goats can be poor in a society in which edible animals are wealth. All that the nomads must buy is the grain grown by their more settled brethren, who will always take a goat or sheep in payment.

The herders can make do with a small wooden sleeping brace used as a pillow, a spear or long knife, and a cloak. The women use a minimum of pots and pans. It is barely an Iron Age existence, but it has succeeded for generations and still succeeds now, when nature and humankind let it be.

It was among these people that the last smallpox battle would be fought.

While the visible disease was being fought in Ethiopia, the virus was at work among the nomads, using its ability to sustain itself in small populations, to survive the desert, hitching a ride with one band and infecting just enough people to stay alive until it could contact another band and achieve viral immortality.

Somalia had been free of the endemic disease since 1963. There were occasional importations from Ethiopia and Kenya, almost always among the nomad population, but these outbreaks were contained easily, even with the minimal efforts of the government and WHO, which had been active in Somalia since 1968. Again the disease was variola minor, and it was the least of the worries of a people frequently hit by tuberculosis, malaria, malnutrition, anemia and snail fever (schistosomiasis). Even when WHO became involved in Somalia it did so with only one epidemiologist, a Pakistani.

Shortly after the "last case" was found in Ethiopia in 1976, there were reports of smallpox in the Somali capital of Mogadishu. It was later learned that the disease had been discovered a few months before, in April or May, but was suppressed by minor government officials. Because of the suppression, containment measures were instituted too late.

At the encouragement of the WHO representative, the

government instituted a series of nighttime searches in the city, looking for more cases.

"When people went to bed at midnight, we would start our searching, entering every house to see if somebody was being concealed," says Dr. Abdullahi Deria, the assistant minister of health. "Four times it was done, something like three thousand people were involved, the army, police, volunteers. It was a big operation. I took part in the last one. Up all that night, door-to-door, the entire city, divided into sectors."

WHO shipped in a couple of international epidemiologists to help. The government reported thirty-nine cases, but probably there were many more that were never written down. The primary contact, the source of the disease, never actually was traced. The person was believed publicly to be a nomad. Privately it was thought that one of the guerrillas in the Ogaden brought the disease back to Somalia with him.

WHO, in the country on its own request more than on the Somalis', didn't press to ask. It was disturbed by this outbreak because Somalia was the last place in the world with smallpox and the eradication program would fail if the disease was not controlled and wiped out there.

Whether the outbreak was isolated and unconnected to later occurrences or was just their first stage is not really known.

WHO called a meeting in Nairobi, inviting the Somali Republic, Kenya, Sudan and Ethiopia. Ethiopia and Somalia were at war in the Ogaden. Somalia and Kenya had strained relations because of disputed border territory. Yet the adversaries met in Nairobi and cooperated.

The four countries agreed to coordinate efforts. Sudan was smallpox free. Kenya had reported its last case (a Somali nomad) in February. Ethiopia had no cases, and Somalia had just gone through its little scare. The countries wanted to make sure that the surveillance was stepped up for six months and that any outbreaks were prevented from crossing borders. Why should warring or hostile nations feel that strongly?

"First of all," Deria says, "our government decided to get rid of smallpox. That was the main thing, with global implications: it's a worldwide policy, and therefore everybody was anxious to play his part. Since the area is the last area where smallpox appears, I think everybody realizes we would be holding back the whole world if we did not do this.

"Economically it is now realized that it is logical to get rid of a disease rather than just control it. . . . In our minds we associate smallpox with death and devastation and definitely, I think, that plays a part in it," Deria says. "But I think, more than anything else, I think it is a question of policy to get rid of the disease."

The Nairobi meeting ended March 15. The next day the health ministry heard a rumor of smallpox in the Bakol region, bordering Ethiopia, and sent an officer down to take a look. On the 18th he reported that there was smallpox in the south.

The disease had been festering in four camps set up earlier that month by nomads in a relatively well-populated, predominantly agricultural area, and had spread out. Specimens were collected and sent to Geneva, which passed them on to CDC. James Nakano's lab in Atlanta verified what the clinicians already knew: it was unquestionably smallpox.

WHO had been so sure that the smallpox outbreak in Mogadishu was the last that it had scheduled an announcement of The Last Case and had even booked a press-conference satellite relay between Washington, New York, Geneva and Mogadishu. Now it quickly canceled the announcement and called in the troops.

The eradication program in Geneva was currently in other hands. D. A. Henderson had resigned his former posts to take a position as dean of the School of Public Health at

Johns Hopkins University in Baltimore. He had accomplished what he had set out to do, for all practical purposes. When he started, the disease was endemic on three continents. Now the concern was a small group of nomads in the Somali desert. Variola major had been vanquished. Many of the world's nations had stopped manning airports and seaports to inspect vaccination cards. Some had even stopped vaccinating altogether. The developed nations were already saving millions of dollars. WHO was maintaining only what seemed to be mop-up operations and had a well-oiled, efficient machine to deal with them, so Henderson decided to move on. He left the terminal phase of the worldwide program in the hands of his WHO assistant, Isao Arita. Arita had the difficult job of keeping the program going after some of the passion seemed spent.

Zdeno Jezek, somewhat idle and bored since the last outbreak in India, was sent in from New Delhi to head the program. CDC's Stan Foster was shipped in later as it became apparent that this outbreak was not small. Vaccinators like Alan Schnur and administrators like John Wickett also were dispatched to the scene.

Arita himself popped in for a visit. He was concerned about the gap in knowledge that led to the Mogadishu surprise. Somehow a chain of infection had eluded the surveillance teams: this was not supposed to happen.

"We thought we would be able to link the Ethiopian outbreak and the Mogadishu outbreak," he says. "But despite all efforts we couldn't find out the source of the infection in Mogadishu." WHO also was missing any possible link between Mogadishu and the new episode in the south.

The last case in Kenya also was suspicious. It was an outbreak brought into the Mandera district of northern Kenya by a Somali, foci unknown, but either Ethiopian or Somalian. "Again, we couldn't identify the index case [first case] for this outbreak," Arita says. There were five cases in Kenya.

This importation had led to the March Nairobi meeting, and now it was more important than ever to press the six-month extensive surveillance efforts pledged by the four countries at their Nairobi meeting.

WHO decided to move in force to squelch the outbreaks. It was decided that the Somalis should declare a state of emergency so that the other U. N. agencies and other countries would be brought into the program.

"I telegraphed Mogadishu telling them to send me a telegram in return declaring the country a disaster area so that I could fly in Land-Rovers," John Wickett says. "That was on the 17th of May.

"It's a game you play: Go after normal international funding or after disaster funds. Normal international assistance in WHO is multilateral or bilateral. This takes a long time because you go through the whole business where you get a project, and you submit it to WHO. And of course the country you are trying to hit up for money, they go back to their own people in the country and check it out, and it goes on and on. But there is, for most countries, a disaster fund. The trick is to hit the disaster fund and that's what we had to do in Somalia. So why the hell not declare smallpox in Somalia a disaster?" The substance of Wickett's telegram was that he needed to get the Land-Rovers to Somalia before the disease got completely out of control and all of East Africa was reinfected.

Wickett explains what led to his scheme and the steps he took.

"We found out that you can get UNICEF Land-Rovers and certain emergency supplies," he says. The Land-Rovers were in Copenhagen. Wickett remembered that U.S. military pilots have to put in a certain number of training hours to keep their proficiency rating.

"So I phoned the U.S. Mission in Geneva and said, 'Hi, why don't you divert a couple of training flights and put a couple of my Land-Rovers on board and fly them down to

Somalia? Like tomorrow?' They said, 'Well, why don't you get in touch with UNDRO [United Nations Disaster Relief Organization] in the Palais des Nations?' I said, 'O.K.,' and I phone them, and UNDRO said, 'Why don't you get in touch with your emergency relief operations in WHO?' So I did that, and the guy said, 'Why don't you have Somalia send a cable to UNDRO, declaring themselves a disaster?' So I said, 'All right, I will.' All this took four hours, and I sent a cable."

The next day Somalia sent the appropriate cable back. It took ten days for UNDRO to react.

Finally all the bureaucracies in Geneva got together and on May 27 UNDRO put out a formal appeal for disaster help. Within twenty-four hours Britain had four Land-Rovers aboard a stretched Hercules on its way to Somalia. The Canadian armed forces sent two Hercules propjets (on supposed training missions) to Copenhagen to pick up the UNICEF Land-Rovers.

The vehicles were precleared by customs. The planes landed at the little seaside airport at Mogadishu, opened their doors, dropped out the vehicles and took off again. The longest turnaround time was seven minutes. One Canadian plane did it in four. "Touch down, drop the ramp, run it up and whew," Wickett says, "they're gone." They left three Land-Rovers sitting on the runway for smallpox.

There were other airlifts; some worked well, some not.

Back in Mogadishu the situation looked glum. Jezek, usually imperturbable, was quite worried.

"For eight weeks I wasn't able to sleep," he says, "not that I was overworked, but they were telling me that there were fifteen outbreaks very close to Kenya." The Kenyan program was not very efficient in the villages, and if the disease was drifting across the border it would take a year for the Kenyans to control it. The ten-year extermination goal set by the World Health Assembly was about to run out, and he could envision the program bogging down into years of puttering around the Horn of Africa chasing nomads with variola minor.

Only once before did he lose sleep over the program, Jezek says: in India, when he knew that the entire concept of global eradication depended on success there. Now he was faced with another difficult period, having the responsibility for the last eradication program and not knowing where the disease was coming from or how widespread it was.

Jezek is a graduate of Prague's venerable Charles University and spent his early years as a public health officer in Czechoslovakia, working under Karel Raska, D. A. Henderson's early champion in WHO. Jezek has an M.D. and also a Ph.D. in epidemiology and microbiology; he inaugurated the first smallpox program in Czechoslovakia. In 1964 he went to work for WHO, beginning his life of roaming and adventure. Toward the end of the Indian campaign, the Czech government tried to recall him to Prague. Only the intervention of Grasset with the Czech ambassador in India prevented his departure. She said she would rather have Jezek than a million dollars.

Jezek's tour in Somalia was supposed to be brief and had been depicted as fairly easy.

"At the time they told us, 'Go there for a short trip and finish work on the outbreaks.' We were thinking the same way," he says. "For the first week I will be busy but what I will do the next week, I don't know. But after we came we found that the situation was much more serious. I remember our first trip in our area, during our whole day, I was able to find ten outbreaks and my colleague another fifteen. In the whole day we were not able to find one who was protected, who was vaccinated."

Most of the program had been centered around Mogadishu and the outbreak was now in the south, so everything had to be shifted.

"People were mass-vaccination oriented," he says. "Then they thought of beginning thorough investigation of the foci, tracing the contact, backward tracing from the source, forward tracing for other possible cases. Therefore the first containment activities were very slowly introduced and many direct contacts remained unvaccinated.

"We started with the surveillance-and-containment orientation in May, 1977."

Even as late as that May, when the first outbreak in Mogadishu supposedly had been ended, Jezek was able to find cases clustered along the three main highways leading toward Kenya and Ethiopia.

"The possibilities of the program were so limited that we were probably only dreaming about making surveillance outside Mogadishu. At the time I had thirty vaccinators and one adviser. The program was very limited."

But gradually, as more people were shifted in from Asia, the professionalism of the WHO efforts improved. It was a group of people who knew what to do, Jezek says, "who knew the diagnosis and who were coming now with a little more, shall we say, global approach.

"What does it mean, one focus? Let's take the whole region, how many foci? How many people do we need? How many cars, and so on," he says, imitating the thought processes of the internationals. "Our calculations started. Where are the cars? Cars were not there. Import the cars! We got them."

Finding cases was no problem. A WHO smallpox worker, Abdul Gadir el Sid of Sudan, went to a market one day and saw a case of smallpox. He went back to his vehicle, told his driver that he would take the wheel and deliberately ran into a foot and a half of mud. Forty people came out to push; five of them had smallpox.

WHO had to plug into the health system of the Somali Republic. The arrangement with the government was that WHO would pay all expenses. The agency was so determined to succeed that it sent whatever Jezek and his staff requested. The government, however, would send him people only.

"Gradually we were able to take and train the staff, and I must say that the ministry of public health was able to give us one supervisor for every region. It assigned us thirty-four, not medical officers, but sanitarians who had some health training," he says.

The health system of the country was not elaborate. It had been built from scratch.

Somalia once consisted of three colonies: Italian Somaliland in the south, and British Somaliland and French Somaliland in the north. In 1960 the Italian and British colonies were set free and joined to form the Somali Republic. French Somaliland is now the Republic of Djibouti.

In the Somali Republic the North African desert meets the mid-African savannas. The population is Hamitic and Muslim. Traditionally identified as the land of frankincense and myrrh, its limited pastoral and agricultural economy produces few exports, notably bananas.

When colonial government ceased, the new country underwent political upheavals. A democratic government led by President Abdirashid Ali Shermarke was overthrown by his assassination in 1969 and replaced by a military junta under Major General Muhammad Siad Barre. The new government was not democratic but it was credited with being much less corrupt and more responsive to public needs than its predecessor.

The Siad government, however, was willing to go only so far to accommodate WHO. It was also hampered by the fact that the health delivery system in Somalia is quite limited and is all state-run. There is no private sector. Only a few physicians remained after independence. The others, like Deria, were trained outside the country, usually in Eastern Europe or Italy. (The Somali Republic recently opened its own medical school.) The minister of health, an army colonel and transportation expert, Muse Rabile Good, has little medical knowledge but is acknowledged to know when to listen to experts. Deria and one other physician only are on the ministry staff, and Rabile generally lets them do their work without irrelevant interference. As a member of the central committee of the Somali Revolutionary Socialist Party, Rabile has the power to get things done in the public-health field.

Building a staff was described by Jezek as "a miracle."

"Our first army, I must say, was an army of the people whom we took and hired locally from the bushmen, the nomads," he says. "They got their few hours of training during the night and the next morning they were placed in the outbreak. You can imagine the difficulty because we were an English-speaking people. Plenty of our Somali counterparts were just at the beginning with English. They were used as translators. What the poor bushmen got (from this) I don't know, but we were able to stop the transmissions. This is the first miracle.

"Plenty of people are speaking about how you must have a delivery system, and you cannot do without a delivery system and that it takes generations to build a delivery system. Our delivery system was done during the night.

"The first weeks were fantastic," he goes on. "We were all out in the field. At ten o'clock at night they came here, the district commissioner with thirty people whom he had selected for us. At eleven we started the training. They finished at five in the morning. At six the car was waiting and our just-trained nomads were going to the villages."

His policy, developed and proven in India, was to simplify the work to the utmost.

During the night training sessions Jezek would hold out his hand and tell the vaccinators that there were five steps preliminary to vaccinating. Pointing to each of his fingers in turn, he started with number one: introduce yourself so that the people won't fear that you are recruiting for the army. Two, show them the recognition card with a picture of a smallpox-infected child and the Somali word *faruga,* (smallpox). Three, ask them if they have seen a case. Four, tell them about the reward. Five, tell them where to report a case.

With the bifurcated needle it took only twenty minutes to train the vaccinators. The rest of the time was spent on the rudiments of epidemiology. The permanent gain to the country is a small army of public-health-trained sanitarians.

"This will continue with another public-health program," Jezek says. "You can see what happened in India. After the [smallpox] program plenty of these people are dismissed."

Jezek's organized monthly searches found cases by the hundreds. Each outbreak was traced and isolated by vaccination. The "people's army" numbered 3,000 at its height.

But there were two problems. The first was that the government was still wedded to mass vaccination and Rabile had to be convinced to change the policy.

"I think that at the beginning he was a little skeptical of what we could do," Jezek says. "He must have been suffering because he had not been informed properly about the scope of the problem. When we started to tell him, he would think: 'Probably you are spreading the disease not controlling it, because yesterday there was nothing, no report, and today I have hundreds.' There were so many infected areas that at first we could not cope. But I think when we started to be effective, he became convinced that our methodology was working."

To help Rabile understand, Jezek used military terms, like "attack," "isolate," "victory," "outflank." Finally Rabile decided to trust the WHO experts against the proponents of mass vaccination who also had his ear.

The second problem was what to do with the cases. Here WHO and the government had to reach an accommodation.

"When I arrived in Mogadishu," Stan Foster says, "there was major chaos: the WHO staff was trying to push the Asian strategy of containment which involved house isolation; and the government was trying to push their traditional strategy, which had been isolation in camps. Both were disasters. Really good smallpox workers had tried to make the Asian strategy and make it work. In a little Somali hut which has visitors coming and going, there's no way this is going to work. The government was perfectly correct in saying that the reason the epidemic was beginning to spread was because WHO isolation methods weren't going to work. But government isolation wasn't working always either."

Deria explains that the camps had two distinct disadvantages. "One was that the people were reluctant to be moved away from their relatives, and that led to concealing cases. The other point was, once you move all the infected cases to one area then all your attention was concentrated on this camp, and there was a tendency to forget about the area where the infection occurred. There may be new cases there and the thing may spread."

Foster, after spending some nights with nomads, decided on a compromise using the thornbush *haro*. Instead of trying to confine the nomad cases in hut isolation or to chase them into internment camps, he created an isolation area close by. He had the women build a special hut, surrounded by its own *haro*, just outside the main barrier.

Guards stood at the entrance to the isolation *haro*. A cook was placed inside and everyone confined was paid 5 shillings * each, enough to buy better food than they were likely to get outside.

WHO gave everyone isolated new clothing on release. People in isolation tried to sneak in relatives so that they too would be paid and get new clothing. Some even pretended to be sick.

---

* One Somali shilling equals sixteen cents.

Generally the compromise scheme worked despite attempts to play the angles. There were one or two serious failures. CDC's Steven Jones reported that he had sent a woman into isolation and put two male guards around her *haro.* A few weeks later the disease began appearing in the encampment. Investigation showed that the guards thought that it was beneath their dignity to go to the well for water, so, naturally, they sent the woman. She stood around, chatting with her friends, spreading the disease.

Because of the strongly socialistic and participatory nature of the government, the field workers had little trouble organizing searches.

"The party organization, the district committee was really good," Foster says. "In other words, if you need a hundred workers tomorrow, you just go see the secretary of the district committee and they'd be there the next morning. You'd train them for a day or two, and when you assigned them to a village thirty or forty miles away, even if there was a vehicle there, they wouldn't think about it, they'd just walk."

One district secretary told his people: "O.K., Dr. Foster said to go to the field for three weeks. If I see you in town before the three weeks you go to jail." He used the women's committee in the town to act as spies and make sure no one sneaked back.

The initial requirement of the government was that 50 percent of the workers had to be women.

"The older women were really good," Foster says. "They would go out and live in a village for six weeks or so and they really did a good job. We had some trouble with the younger ones."

The WHO-Somali teams roamed across the southern half of the country as the number of outbreaks mounted through the summer. The terrain was generally flat and usually accessible by motor vehicle. Alan Schnur recalls several times when nomads were hired to cut down trees so that the

Land-Rovers could pass through forested areas. Sometimes the strong vehicles would just drive over the bushes or try to knock down obstacles by force.

The best searchers were the nomads themselves, because they knew the area and were tough. They were motivated by a substantial reward–200 shillings.

WHO started by offering a single reward for each reported case, but it proved counterproductive. The vaccinators were forgetting the fourth finger of Jezek's five-finger exercise. Some wily nomads, having heard of the reward elsewhere,* did not tell the searchers about cases so that they could claim the money themselves. In retaliation, the searchers began to omit telling nomads who might not know so as to keep the 200 shillings for themselves. WHO solved the problem as in Asia by giving two rewards.

The nomad vaccinators worked in teams. If they stumbled across a case, or found one by tracking down a rumor, one searcher would go back to report to the authorities and the other would remain behind to vaccinate everyone. He would keep with the nomad band for six weeks to see if someone was incubating the disease. Not all the nomads liked being followed, even by their own people. A few vaccinators were left in the lurch, waking one morning to find that their charges had stolen away silently in the night.

Tracking nomads down became something of a game at times because they wanted to keep their water holes secret from competitors and cattle thieves. If you asked a herdsman his next stop he would lie invariably.

A few independent businessmen sprang up to collect the rewards. One free lance collected five 200-shilling rewards for reporting cases. That's a lot of money in Somalia.

---

* Nomads sometimes carry transistor radios. The government played announcements of the reward constantly on the country's only radio network. The nomads listened to the radio because they needed to know where the fighting was in case it blocked their planned moves.

"Our policy was if possible to avoid employing the police and authorities to frighten the people," Deria says. "Of course they were always in the background, and if we had any problem with anything, occasionally we would resort to the police. But we never had a big problem."

Learning the psychology of the Somalis was challenging.

"India eventually was not so difficult," Jezek says. "We were there a long time and learned the psychology of the people. . . . Gradually we developed a program that worked very effectively. Here we were on unknown territory, but, amazingly, the system worked. Within *one* month the number of cases seemed to be responding and the teams to be having some influence in this last rampage by the wild variola virus. They were done in six months." The speed was Jezek's second "miracle."

"Even in India it took us two years to prepare the ground—the delivery system. And the delivery system was able to produce the work," Jezek says. "Here the delivery system was starting to work after *one* month. The first month the people get the routine to find where the outbreak is, there was much more outbreaks than we were able to digest. It took us 2½ months before we started to be quicker than the smallpox process and the smallpox process was going very, very hard for maybe one year before we came—uncontrolled. Therefore we were facing not only the present situation, but something that was developing if not for one year then for at least eight months.

"WHO is responsible for the miracle, first of all because they were able to deliver the transport facilities. People without transport are useless. Second of all, we were able to mobilize all experts which we had, and luckily for us they were free."

Indeed, all the old veterans from around the world, appalled that the virus was still loose, rang up Geneva and volunteered for service.

Isao Arita had his hands full answering volunteers.

"The smallpox-eradication program was unique," Arita says. "So many epidemiologists were interested in the program even though they no longer work in the program, that when they heard about this setback in the program in the Horn of Africa I received many letters 'we can help, why don't you recruit us to go to Somalia?' The Old Boy system worked."

The Somalis, normally distrustful and distant with foreigners, agreed to accept anyone that WHO sent—almost. WHO tactfully avoided sending Russians, knowing that Siad had thrown the Soviets out of his country when he caught them favoring the Ethiopians. Eventually WHO dispatched fifty international epidemiologists to Somalia.

And the cooperative border surveillance survived continuing hostilities and yielded a full exchange of information.

By the time the outbreaks were finally beaten down there had been 3,229 cases.

Victory did not come quietly.

Somewhere in the southern part of the country a group of nomads had been harboring the virus. On October 13 they were discovered by government searchers. Among the victims were two children in one family, a boy of about three and a girl of about four. They were unusually ill for variola minor.

There was no smallpox officer in the area, so the district health officer put the children and their mother into his Land-Rover and drove them to the little banana port of Merka, looking for someone to report to. He did not know where the smallpox office was and drove instead to the pink stucco hospital that sits on a hill overlooking the sea. He knew that there was an isolation camp nearby (some were used in towns) and that the smallpox officer would know where to send the children. He arrived at the hospital at about 5 p.m.

None of the hospital officials was there at that hour. The only person who seemed to know what he was doing was a twenty-three-year-old cook, Ali Maow Maalin. Maow went with the party of four in the vehicle to the isolation camp. The children did not look too sick, he thought.

At the camp the children and their mother left the vehicle and Maow was driven back to the hospital. He was in contact with the victims for about ten to fifteen minutes.

Although Maow had been in the hospital for a few years he had never been vaccinated and no one thought much of the incident that occurred the night of the 13th. The girl died two days later.

Nine days after the contact, while at the hospital, Maow began to feel ill and left work early. For the next three days he remained in bed in the one-room stucco building behind the hospital where he lived. Friends and fellow workers visited him often. On the 25th he was admitted to the hospital. The diagnosis was malaria.

Prior to this he had been suffering from high fever, pain in the joints and vomiting. It was mosquito season, so the diagnosis was not unusual. On the evening of the 26th he developed a rash.

On the 27th the attending physician, knowing that there was some chickenpox in the city and not much else, so diagnosed Maow's case and sent him home until he got over it.

Maow, a tall, quiet young man with a broader than usual face and dark brown skin, had seen smallpox posters in the hospital and had seen some victims. He knew that his rash was not chickenpox. He was afraid of reporting himself because he did not want to go to the isolation camp.

On the 30th a friend, a male nurse, visited Maow and reported him to the hospital officials as a smallpox case. (The nurse collected the reward. Maow has received nothing for his misadventure.)

The next day, the 31st, a WHO epidemiologist, K. Markvart, a Czech, confirmed the diagnosis and reported to Mogadishu. That afternoon two guards were placed in front of Maow's door.

At the hospital pandemonium broke loose. The three days Maow had been there under the malaria misdiagnosis were spent in a thirty-two-bed ward. Each patient had visitors. There was also the hospital staff to consider.

The hospital was quickly closed for new admissions to all but emergency cases and no patients were discharged. All former patients were traced. No one could step out of the hospital who was not searched for a smallpox vaccination scar.

By evening vaccinators spread out all across Merka, a town of about 25,000. WHO expected many secondary cases. Merka is a central gathering point for the district and is connected to six other district towns by a bus service. It is only 130 kilometers from Mogadishu.

Mogadishu sent twenty-three more vaccination teams, twenty-five policemen, seven militiamen and nine supervisors.

On November 2 Maow was sent to the isolation camp. About forty people were there. He says that he had already started feeling better. A militiaman and a policeman acted as his house guards.

WHO found 161 possible contacts (see first table).

VACCINATION STATUS	DEGREE OF EXPOSURE		
	FACE-TO-FACE	INCIDENTAL	TOTAL
Unprotected	33	8	41
Protected *	58	62	120
Total	91	70	161

* Protected is defined as having received a vaccination within the last three years.

Of the face-to-face contacts, only twelve had no vaccination scars and were totally unprotected. The others had been vaccinated some time in the past. All were vaccinated and all were visited four to six times to make sure that none of them was incubating the disease.

The vaccinators in Merka worked mainly at night, concentrating first on the fifty houses around the hospital and then spreading out all across the town. It took a week. No one was missed. (See second table.)

AREA	PERIOD	NO. OF HOUSES	ESTIMATED PEOPLE	PERSONS VACCINATED
Horseed	10/31–11/2	792	5,000	3,558
Wadajir	11/3–11/6	738	4,300	2,873
Wadak	11/7–11/13	1,007	10,502	8,092
Checkpoints	10/31–11/14			40,254
Total		2,537	19,802	54,777

The checkpoints were set up on all the roads leading into Merka and were manned for six weeks. All vehicles were stopped. Everyone who could not prove that he had been vaccinated recently was put to the bifurcated needle. Two freighters tied up offshore were not permitted to send their crews onshore until they could come up with valid vaccination certificates or submitted to the needle.

The reward for reporting cases was widely announced at meetings.

For months the whole region was searched for the secondary cases WHO expected. There were 927 village searches in Merka District and 5,984 in Lower Shabelli, where the nomads had been before the outbreak. There were 2,076 visits to nomad camps. The WHO-government teams searched and re-searched houses in both districts, tallying 559,668 visits. There were 5,764 searches in schools. All of this was between November of 1977 and the next March.

After twenty-six days in the isolation camp Maow was

sent home. Six months later his scars were beginning to fade. He is still working in the hospital.

The viruses that made him sick, the last variola free in the world, fell to the ground somewhere in the Merka District, Somalia. They are now extinct in nature.

Ali Maow Maalin is believed to be the last human to suffer smallpox from a natural infection. Ever. WHO missed its ten-year deadline by ten months.

Unfortunately the danger is not over.

Epilogue

WHO and the various governments are still scouring the world for smallpox. There is a $1,000 reward for the first person who reports a confirmed case of naturally acquired smallpox anywhere in the world. Arita, who is now running the smallpox unit, does not expect to have to pay it.

How certain can we be that the disease has been eradicated?

The last natural case (as of this writing) was recorded on October 26, 1977—that of Maow in Merka. Despite door-to-door searches in a half dozen nations there has not been another confirmed case. As time goes by the chances of ever finding another diminish drastically.

In May, 1978, when the question of eradication was put to various officials at WHO headquarters in Geneva, the answers were the same. John Wickett, perhaps, said it best:

"We can be sure because we do surveillance," Wickett

said. "It's a systematic covering of an entire country. And given the characteristics of the disease—it has to go person to person—consider that almost seven months have gone by now since we had a known case. If we were to find a case today it would mean that there would have had to be seven months of ongoing transmission. The incubation period is two weeks, so it means that you would have had to have a case every two weeks, minimally, for the past seven months. That would be a lot of cases. Smallpox does get notified, it's a nasty disease. So it seems very unlikely—the chances that it's present and we haven't found it keep going down exponentially, because you have this continuous chain."

Thus, the more time that goes by without a new reported case, the less likely it becomes that there will be one, because the larger the chain of infection that WHO would have had to miss. Arita's smallpox unit firmly believes that no chain that size could go unnoticed, even in the rugged Ethiopian highlands or the Ogaden.

Certain formalities, however, must be met before WHO can proclaim the disease eradicated. Two years must elapse before a country can be declared smallpox-free and that cannot happen until October 26, 1979, at the earliest for Somalia if Maow is the last case found there.

During those two years WHO and the government must continue extensive and earnest searches to convince an international investigating commission that there are no cases. In Somalia 1,500 rumors a week are being tracked down. Every case of rash and fever is investigated and specimens are flown either to James Nakano in Atlanta or to his Russian counterpart, Dr. Svetlana Marennikova in Moscow.

Two years has been selected because that is far longer than any past outbreak has remained hidden. In only four countries have there been outbreaks reported after WHO felt that it had broken the chains of infection: Botswana, Brazil, Indonesia and Nigeria. It happened three times in

Botswana. At no time did the outbreak remain unreported to WHO for more than thirty-four weeks, so two years is three times longer than reported in any earlier incident.

During this time, the special searches by Arita's teams are continuing in previously endemic countries, all based on several premises:

1. Smallpox is a notorious disease, and if there is a case anywhere, it would sooner or later be reported to authorities, particularly in countries which have had extensive eradication programs and rewards. WHO keeps watching health records in the previously infected countries to see if any smallpox has been reported to the governments but not passed on to WHO.

2. Any outbreaks would most likely be found in the vicinity of the last ones known. Searchers go over the area of the last outbreaks looking for cases and asking the residents to report any rashes and fever.

3. Smallpox leaves scars. WHO searchers perform scar surveys, looking for people who seem to have been scarred recently. They are particularly seeking out children born since the last known case. If any of them have scars, the searchers can assume that the disease was around.

4. In areas where chickenpox is common special care is taken to make sure that all cases are diagnosed correctly. Samples are sent to Atlanta and Moscow.

A global commission appointed by WHO's director-general is now overseeing the work in Asia and in the Horn of Africa. If it confirms what the field people tell it, the World Health Assembly will be informed and in its 1980 meeting (if no further cases are found) the assembly will declare the world free of the disease.

Although WHO would be bitterly disappointed if another case were discovered, its victory would be diminished only slightly. To worry about a half dozen cases of variola minor in the Ogaden desert when once millions died of the disease each year seems almost ludicrous. Yet the eradicators are so determined to have a complete victory that the news of another case, and the potential spread, would be a harsh blow.

The next question is whether or not the eradication effort was really worth the cost? It seems obvious that the answer is yes, but it may not be clear who was doing whom a favor in eradicating smallpox.

Six hundred WHO staff, over ten years, permanent or part-time, from fifty-one countries, participated in the program.

WHO received monetary and practical aid from forty-four countries. From 1967, when the expanded program began in earnest, through 1978, that aid will have totaled $96.5 million, an average of $8 million a year. Donations came from various sources, direct and indirect. The regular WHO budget provided $34.3 million, the United States $23.3 million, the Soviet Union $13.4 million, Sweden $13.3 million. Other large contributors ($50,000 or more) included Canada, Denmark, the Federal Republic of Germany, India, Iran, Japan, the Netherlands, Norway and the United Nations Emergency Operation. By and large the donations all came from the developed countries.

The developing nations, those where the disease was still endemic when the program began, spent about twice the money given by the developed nations, something like $200 million. In almost every instance the cost was not very much more than they were spending already each year for smallpox control.

If one can judge the eradication of a disease by businesslike cost-benefit studies, the smallpox program comes out looking like a miracle. It cost the United States $150 million a year for vaccination and quarantine activities just to prevent smallpox from sneaking across its borders. For a fraction of that, in support of WHO, the expensive machinery has been taken down. WHO estimates that every year $1 billion was spent around the world for smallpox control. In 1962 Pakistani travelers brought smallpox to the United Kingdom, causing sixty-seven cases and requiring the vaccination of 5.5 million people. It cost the British government $3.6 million to handle that one outbreak. Sweden spent $750,000 to fight a twenty-seven-case outbreak caused when one sailor brought the disease in from Asia in 1963. No one knows what it cost the Yugoslav government to battle the 1972 outbreak.

So for a third of $1 billion, spread over twelve years, money need no longer be spent on smallpox and can be used on other health problems. It was a major investment, but it has already paid for itself many times over.

The nations that benefited the most were the developed nations. In India and Bangladesh it was only one of a great many horrors to afflict the people and not that big an economic drain. The director-general of WHO has stated that the eradication program was a $2 billion gift from the poor nations to the rich, because it was the latter that benefited the most, at least in financial terms.

Can the disease come back again?

The answer, unfortunately, is yes. But only with human help. Scientists are pretty sure that there is no animal reservoir, so it is not lurking in some jungle or desert. In those countries which wiped out the disease earliest, every single subsequent outbreak was traced to a traveler from abroad.

There is the possibility, however, that another virus disease will mutate and take over the ecological niche left empty by variola. Scientists are already concerned about one such virus, monkeypox.

About thirty confirmed cases of monkeypox virus infecting man have occurred since 1970. The disease was transmitted from an animal in all but two cases where it was transmitted from one person to another. At this transmission rate the chances of a sustained chain of infection seem slim unless the virus mutates to become more virulent.

The monkeypox cases were all in west Africa, most of them in Zaïre, so WHO is watching that area very carefully.

So where is the danger? Laboratories.

At this writing there are about eight laboratories * in the world with variola viruses locked and frozen for study. They are kept for scientific curiosity and to be used to help identify any suspicious specimens shipped for study. WHO wants to cut those labs down to the two or three most likely to handle the viruses with care: labs in Atlanta, Moscow and London. The World Health Assembly has asked the other laboratories still holding the virus either to destroy it or ship it to a central cooperating lab.

But for various reasons labs have refused. WHO's fears are deep and well-founded.

"If smallpox returns to devastate mankind it will come from a laboratory," says Arita.

There is now one laboratory in the United Kingdom keeping smallpox samples in its freezer, St. Mary's Hospital in London. Once there were three.

On February 28, 1973, a young Irish woman, Ann Algeo, a laboratory assistant, visited the pox laboratory at the London School of Hygiene and Tropical Medicine. She went to the lab frequently to use the equipment there.

"It is hard to generalize," she said later, "but my visits

* Statements from mainland China refer to "laboratories" in the plural, so no one knows just how many labs with smallpox virus are there.

would have lasted anything from a few minutes to perhaps three or four hours.

"While in the pox virus laboratory I was not working with the pox viruses and certainly never touched anything connected with them. I can remember seeing eggs being inoculated a number of times—and on perhaps two occasions eggs were being harvested.

"This was carried out on an open bench at the opposite side of the laboratory to where I worked. I remember inquiring how it was done and for a very short period watching."

The eggs were used to grow viruses. Laboratory viruses lose some of their potency if they are not activated and made to reproduce once in a while. Chick membranes are used for this purpose and to grow viruses for identification.

The laboratory, like many in Britain, is quite old. It is also quite small, so small in fact that the viruses were kept in a freezer in the hallway. A professor at London University, Reginald Shooter, had once been asked to evaluate the lab. What he found disturbed him.

"It was grossly overful with equipment and with items stored on shelves. The working surfaces are of wood and are in bad condition," he said.

"The laboratory is also very dirty, the perhaps not surprising result of a deliberate policy to clear it out once a year." The practice was designed to minimize endangering the janitors.

"I gathered that either old coats were worn and then incinerated, or other gowns used that were then washed in a sink in the laboratory and then dripped dry."

The people who worked in the lab thought that it was safe and knew how cramped things were at the school. "If it was too small to do smallpox work, I would not have done so," said Dr. Charles Rondle, head of the lab.

"Space at the school is at a premium. I had several times

said that if a laboratory fell vacant I would be grateful to use it. I was badly in need of another laboratory."

The refrigerator was in the hallway, he said, because "at that time we had nowhere else to put it."

Only once had anyone asked Miss Algeo if she had ever been vaccinated. She had, in fact, been vaccinated about a year earlier but it did not take very well.

On March 11 Miss Algeo became ill. She ran a fever and had pain in her joints. Two days later she visited her doctor, John West, who assumed that she had the flu and treated her accordingly. On March 15 she developed a slight rash and began to vomit. She was admitted to St. Mary's Hospital (tentative diagnosis: glandular fever) and placed in a general ward. In the bed next to her was Mrs. Norah Hurley, who was visited frequently by her son, Thomas, and her daughter-in-law, Margaret. On March 20 Mrs. Hurley was released from the hospital and another patient took her bed.

Doctors still were not sure what Miss Algeo had. Because she worked in a fungus lab, it was suspected that she might have picked up some arcane infection. On March 22 she was visited by a friend, Christine Philpot, thirty-eight, a mycologist. "I thought it would be kind to go see her," she said. At the request of Dr. Donald Mackenzie, head of the mycology lab, Miss Philpot took some skin scrapings from Miss Algeo. "I took them without obtaining any permission from a doctor or nurse," she said.

She brought the scrapings back to the mycology lab and, with a doctor, looked at them through an electron microscope. The two saw some brick-shaped objects in the samples.

"I did not do anything about this development at the time, as it did not occur to me that this finding had any significance. Dr. Ellis [the other person with her] said they looked like pox virus. I had no idea what kind of virus."

But the next day Mackenzie came and looked at the viruses. He knew immediately what they were—variola.

Miss Algeo was taken immediately to Long Reach Isolation Hospital in Dartford, Kent. The diagnosis of smallpox was confirmed that day.

Investigators went to St. Mary's and tried to get a list of all the contacts Miss Algeo might have had while she was there. Because she had been released, Mrs. Hurley's name was not placed on the list by the harried nurses. Everybody else was contacted, vaccinated and watched.

On March 31 Mrs. Hurley's son and daughter-in-law became ill. They thought that they had food poisoning and their doctor agreed. But they got worse and the next evening they were admitted to Herndon Isolation Hospital. The diagnosis of smallpox came three days later, and they too were transferred to Long Reach.

On April 6 Margaret Hurley, twenty-nine, died. Her husband was still seriously ill. Their house in Middlesex was sealed off. The Hurleys' two children were sent to their grandmother in Harrow. A relative and a boarder in the Hurley house were placed in quarantine.

The announcement that smallpox was loose set off minor hysteria. A number of countries required vaccination of travelers from Britain. Medical teams at Heathrow Airport vaccinated four thousand people a day so that they could travel abroad. Every possible contact of the Hurleys was tracked down and vaccinated.

In Bedfordshire, near the Hurley home, seven people believed to be contacts were also placed in quarantine. Government personnel sent them food and checked them every day to see if they came down with the disease.

Dr. Melville MacLeod, the local health officer said, "At present this is not a situation where we feel mass public vaccination is necessary." Nonetheless, many people lined up at health centers for vaccination.

Accusations of negligence on the part of both St. Mary's and the School of Hygiene were reported in the press. The institutions were obliged to defend themselves publicly.

On April 15 Thomas Hurley, thirty-four, died of smallpox. A nurse who treated the Hurleys when they were at Herndon became ill and was admitted to Long Reach. "We do not know yet if she has the disease," a spokesman said, "but obviously there must now be tracing along the line to see whom she has come in contact with." It turned out that she did have very mild smallpox.

Vaccinations were continuing at London's two airports. A British family of four was made to leave the Spanish island of Minorca in the Mediterranean when it was found that the four-year-old daughter had not been vaccinated. She was suffering from a skin disease and could not be vaccinated, but the authorities on Minorca did not want her there.

The British government ordered a public inquiry into the outbreak. Much of the testimony centered around the physical condition of the laboratory.

"Quite frankly I was appalled at the conditions," Dr. Henry Darlow of the government's research establishment said. He had visited the lab two years earlier and had been struck by several unsafe conditions.

"There was glassware all over the bench, and I said it looked a bit like a medieval alchemist's workshop. I expressed horror that people should be expected to handle pathogens in so mean a laboratory." He also noted that there was no safety cabinet for handling dangerous viruses. Conditions there had not changed since his earlier visit, he said.

Part of the blame for the outbreak was placed on the failure of the doctors to diagnose the smallpox more promptly. One doctor said that he didn't believe the smallpox diagnosis when it was made because no one seemed very alarmed.

"I felt even more reassured as time went by. I thought if it was likely to be a case of smallpox that the experts would be there instantly, but the longer they took to come the less likely it would be," the doctor said. It took the experts several hours to show up at St. Mary's to look at Miss Algeo (who survived the infection). The doctors did not know that she had been in a pox virus lab until after that initial examination.

The committee of inquiry report brought new safety measures to the School of Hygiene. The school set up a safety board to formulate and enforce a code. Everyone who came near the lab in the course of his business would be vaccinated, and the government set up a special group of epidemiologists to control any possible subsequent outbreak. The school eventually gave up its collection of viruses.

The London accident was not the last. It was Birmingham's turn next.

On August 11, 1978, ten months after the last case was reported in Somalia, Mrs. Janet Parker, a laboratory photographer, fell ill. Her office at the Birmingham University School of Medicine was on the floor above the microbiology lab where samples of variola are kept. She stayed home because of the illness and was not concerned until the 19th, when she broke out in the telltale rash. She was admitted to Catherine Barnes Hospital in East Birmingham under strict isolation. A few days later a diagnosis of smallpox was confirmed.

Mrs. Parker had been vaccinated in 1966 but she received no revaccination because her work did not involve coming in contact with variola.

The wing of the medical school with the lab was sealed off as a precaution. WHO flew in a consultant to observe the operation. Some two hundred contacts were identified and thousands of people in Britain's second largest city were vaccinated.

On September 2 the forty-nine-year-old Professor Henry Bedson who ran the laboratory was found with his throat cut in an apparent suicide attempt. He had left a note asking forgiveness from those who trusted him. He had been reported under "considerable strain."

Bedson was placed on a respirator to keep him alive, but five days later, with no brain waves evident on the recording instrument, the doctors pulled the plug and let him die.

Meanwhile Mrs. Parker's condition was deteriorating rapidly. On September 11, eighteen days after the smallpox diagnosis, Janet Parker died.

The tragedy was not over. Frederick Whitcomb, Mrs. Parker's seventy-one-year-old father, admitted to the hospital as a possible contact, died of a heart attack. Her mother, seventy, was also diagnosed as having smallpox but recovered.

Preliminary reports indicate that the virus that infected Mrs. Parker might have come from the laboratory. In fact, there is no place else it could have come from. Mrs. Parker had not been away from England in the year before her fatal illness and outside of the labs in Britain, variola does not exist in the country.

Both accidents in England could have been calamitous. Most people in the Western world do not have full protection against smallpox, and, as both Mrs. Parker and Miss Algeo demonstrated, having been vaccinated once does not guarantee lifelong immunity to the disease. If variola ever escaped from a lab in quantity, or if someone incubating the disease after an accidental infection met many unprotected people, there could be a serious and very deadly outbreak. That's why Arita and WHO are concerned.

Three labs in the United States hold smallpox virus: Nakano's CDC laboratory; the American Type Culture Collection, a medical archive in Rockville, Maryland; the U.S. Army Medical Research Institute of Infectious Diseases (USAMRIID) at Walter Reed Army Hospital in Bethesda, Maryland.

Why does the Army need smallpox virus?

"The only reason to have smallpox virus is for offensive [biologic warfare] purposes," John H. Richardson, director of biosafety at CDC, said in an interview with *Science* magazine. The Army maintains that it needs the virus as a standard in case someone else unleashes it in warfare. One expert also suggested that "USAMRIID maintains stocks in the event that at some future time they can no longer rely on CDC." He means that the Army is afraid that CDC will lose or misplace the viruses.

The American Type Culture Collection board has voted to keep the virus although it will not lend out samples as it does with other viruses and bacteria. It also is afraid that CDC will lose its samples.

"I personally feel that ATCC is probably the best place in the world for preserving things of this sort," says Adrian Chapell, of CDC and a former ATCC board member. Richardson does not agree.

"For damned sure, the ATCC storage area does not meet the recommended WHO standards for containment of smallpox." *

And finally, there is one more question: Can another disease be eradicated using the kind of organization and dedication WHO put together for smallpox? The answer is not clear.

WHO itself seems to have backed away from any further projects of this nature. The chafing from the smallpox team is still sore. Vertical programs are much too disruptive to

* In January, 1979, ATCC and the Army agreed to give up their stocks.

that lovely green hillside in Geneva, and too many hallowed rules and philosophies were twisted or bent. Too many new and controversial precedents were set.

It may be that WHO will proceed in the future as normally as it did in the past, using the regular bureaucracy, established machinery and careful rules. WHO has now gone into a program of extensive immunization against childhood diseases (under Rafe Henderson)—a worthy undertaking, but one that will never succeed in reaching all the children that need it, because the usual way of doing business at WHO is too slow and inefficient. Another plan to improve primary health care will fare no better.

The malaria program is in such disarray that the staff assigned to the eradication organization has been cut to a fraction of its former size and moved to an annex building, almost as if WHO were embarrassed to have it around.

And there is still much bitterness left from all the squabbles. When the smallpox program wound down Nicole Grasset decided that her next project would be preventing childhood blindness. WHO refused to give her a job.

Grasset, the Brilliants and some of the other smallpox workers have formed their own foundation, called SEVA, after the Sanskrit word for "service to humanity." They are now engaged in raising money for Grasset's work in Nepal on preventable blindness, and for a hospital in India. Brilliant has enlisted some of his friends from his hippie days to raise the money. WHO may provide a grant.

And, finally, there are very few diseases that are as vulnerable as smallpox. Most have animal reservoirs that make eradication unlikely. A malaria vaccine is about ten years off. No leprosy vaccine exists. Measles possibly can be eradicated but it is not serious enough to justify a worldwide campaign, although its eradication in such highly susceptible areas as parts of Africa would be desirable.

But the lessons learned in the smallpox campaign are still valid—and the idealism that fired the workers is still alive.

Can it be done again?

"The human type still exists," Nedd Willard says. "Some of the smallpox people went into or came out of leprosy, which is not a gay disease to work with. So there is a fund of idealism in the world that can be tapped. Hell, they had more volunteers than they knew what to do with!"

"I think so," says Bill Foege. "The payoff has to be big enough, has to be important enough for people to dedicate themselves like that."

"I keep looking around and repeatedly see people come up with that kind of dedication, whether it's to a political figure or to make money in the stock market; it says to me that there are many other areas in health where we could replicate this spirit."

Larry Brilliant is the most optimistic. Brilliant now teaches at the University of Michigan School of Public Health, helping visiting specialists from less developed nations cope with their problems, using his experience with smallpox as an example. Also he is training young American health workers to go forth into the world and raise holy hell.

"You don't find many smallpox people who want to own Cadillacs, and who want to have fancy mansions," he says, "who want to be selfish about their resources. People have been changed by the smallpox program. They want to take what they can only so that they can give it back tenfold. Each one of us will multiply by the vision we have shared. You have proof now that people can come together, work together in harmony for a common goal.

"We've all been touched by something bigger than ourselves. The arrogance that I felt as a know-it-all radical—or the arrogance that some of my colleagues encountered in some Indian government officials—or the arrogance that some of the Westerners showed toward India—that's all disappeared. We really stand in awe, in all humility, at the prospect of something much bigger than us."

"We were at Nicole's house in Delhi one night," Girija Brilliant says, "and after dinner we started talking about the spiritual reasons for being in the program. What was obvious, as we talked, was that each person in their own [spiritual] way had come to it, and no one looked [down] at the other and said, 'What an odd thing to say.' "

It was poignant when they were finally disbanded, Girija says. But the Brilliants insist that the army is still out there.

It is just waiting to be called upon again.

Sources

Instead of littering the text with footnotes, I have saved the reference materials for the end of the book. Besides those listed, I have used a great many news releases issued by the information office of the World Health Organization and a manuscript by D. A. Henderson written for the 1978 Medical and Health Annual of the *Encyclopaedia Britannica.* All interviews listed in this section were made in person, except where otherwise noted.

Chapter I
The Elegant Killer

INTERVIEWS:
Walt Orenstein, June, 1976; James Nakano, April, 1977;

John Obijeski, April, 1977; Martha Thieme, April, 1977 (all Atlanta).

BOOK:
Locke, David M. *Viruses: The Smallest Enemy.* New York: Crown Publishers, 1974.

UNPUBLISHED MATERIAL:
Nakano, James, "Comparative Diagnosis of Poxvirus Diseases." Manuscript. Center for Disease Control, Atlanta.

Chapter II
The Plague of Athens

BOOKS:
Alivizatos, Gerasimos P. *The Early Smallpox Epidemics in Europe.* Athens: Univ. of Athens, 1950.
Ar-Razi, Abu Bacr Mohammed Ibin Zacariya. *A Treatise on Small Pox and Measles,* transl. William Alex Greenhill. London: Sydenham Society, 1848.
Botsford, George Willis, and Robinson, Charles A. Jr. *Hellenic History,* 4th ed. New York: Macmillan, 1956.
Creighton, Charles. *A History of Epidemics in Britain.* Cambridge: Cambridge Univ. Press, 1891.
Edwardes, Edward J. *A Concise History of Small-Pox and Vaccination in Europe.* London: H. K. Lewis, 1902.
McNeill, William H. *Plagues and Peoples.* Garden City: Doubleday, 1976.
Moore, James. *A History of Smallpox.* London, 1815.
Thorndike, Lynn. *A History of Magic and Experimental Science.* New York: Columbia Univ. Press, 1958.
Thucydides, *The Complete Writings,* transl. Crawley, New York: Modern Library, 1934.

CHAPTER III
OF ALL THE MINISTERS OF DEATH

BOOKS:
Clark, E. G. *The New London Practice of Medicine.* London, 1811.
Creighton, op. cit.
Edwardes, op. cit.
Huxham, John. *An Essay on Fevers.* London, 1794.
McNeill, op. cit.
Sydenham, Thomas. *The Whole Works.* London, 1705.

CHAPTER IV
"LOVING HER AS I DO"

ELIZABETH I

BOOKS:
Luke, Mary M. *Gloriana: The Years of Elizabeth I.* New York: Coward, McCann & Geoghegan, 1973.
Williams, Nelville. *Elizabeth the First, Queen of England.* New York: Dutton, 1968.

MARY II

BOOKS:
Hamilton, Elizabeth. *William's Mary.* New York: Taplinger, 1972.
Traill, H. D. *William the Third.* London: Macmillan, 1892.

LOUIS XV

BOOKS:
Gramont, Sanche de. *Epitaph for Kings.* New York: Putnam's, 1967.

Loomis, Stanley. *Du Barry: A Biography*. Philadelphia: Lippincott, 1959.
Perkins, James Breck. *France Under Louis XV*. Boston: Houghton, Mifflin, 1897.
Williams, H. Noel. *Memoirs of Madame Du Barry*. New York: Collier, 1910.

Chapter V
La Noche Triste

Birney, Hoffman. *Brothers of Doom*. New York: Putnam's, 1942.
The Conquistadors. Edited and transl. by Patricia de Fuentes. New York: Orion, 1963.
Crow, John A. *The Epic of Latin America*. Garden City: Doubleday, 1971.
Duffy, John. *Epidemics in Colonial America*. Baton Rouge: Louisiana State Univ. Press, 1953.
Fehrenbach, T. R. *Fire and Blood*. New York: Macmillan, 1973.
Hemming, John. *The Conquest of the Incas*. New York: Harcourt Brace Jovanovich, 1970.
Marks, Geoffrey, and Beatty, William K. *Epidemics*. New York: Scribner's, 1976.
McNeill, op. cit.
Nicole, Christopher. *The West Indies*. London: Hutchinson, 1965.
Prescott, William Hickling. *The Rise and Decline of the Spanish Empire*, edited by Irwin R. Blacker. New York: Viking, 1963.
Stearn, E. Wagner, and Stearn, Allen E. *The Effects of Smallpox on the Destiny of the Amerindians*. Boston, Bruce Humphries, 1945.
Woodward, Ralph Lee. *Central America, A Nation Divided*. New York: Oxford, 1976.

CHAPTER VI
JENNER

BOOKS:
Dolan, Edward F., Jr. *Jenner and the Miracle of Vaccine.* New York: Dodd, Mead, 1960.
Fisk, Dorothy. *Dr. Jenner of Berkeley.* London: Heinemann, 1959.
Jenner, Edward. *An Inquiry Into the Causes and Effects of the Variolae Vaccinae.* London: Sampson Low, 1800.
Montagu, Lady Mary Wortley. *Selected Letters.* Edited by Robert Halsband. New York: St. Martin's, 1970.
Winslow, Ola Elizabeth. *A Destroying Angel.* Boston: Houghton Mifflin, 1974.

MAGAZINES:
"Immunization Against Smallpox Before Jenner" by William L. Langer, *Scientific American.*
"Smallpox Vaccination: A Swirl of Controversy," by Elise Hancock, *Johns Hopkins Magazine,* Nov. 1976.

CHAPTER VII
THE BELLS OF BOSTON

BOOKS:
Bell, Whitfield J. *The Colonial Physician and Other Essays.* New York: Science History Publications, 1975.
Beebe, Lewis. *Journal.* Philadelphia: Pennsylvania Historical Society, 1935.
Brodie, Fawn M. *Thomas Jefferson, An Intimate History.* New York: Norton, 1974.
Duffy, op. cit.

McNeill, op. cit.
Stearn, op. cit.
Thomason, Adam. *A Discourse on the Preparation of the Body for the Smallpox: And the Manner of Receiving the Infection.* Philadelphia: B. Franklin and D. Hall, 1750.

PAMPHLETS:
"A History of the Natural and Modified Small-Pox; or the Variolas and Varioloid Diseases as they Prevailed in Philadelphia in the Years 1823 and 1824," by John K. Mitchell and John Bell. Philadelphia, 1825.
"An Essay on the Expediency of Inoculation and the Season Most Proper for It," by Lauchlin Macleane, Philadelphia, 1756.

Chapter VIII
The Vaccine Wars

BOOKS:
Bell, Luther M. *An Attempt to Investigate Some Obscure and Undecided Doctrines in Relation to Smallpox, Varioloid and Vaccination.* Boston: Marsh, Capen & Lyons, 1836.
Bell, Whitfield J., op. cit.
Blake, John B. *Benjamin Waterhouse and the Introduction of Vaccination.* Philadelphia: Univ. of Pennsylvania Press, 1957.
Brodie, op. cit.
Buist, John B. *Vaccinia and Variolation: A Study of Their Life History.* London: Churchill, 1887.
History of Amer. Epid., edited by Franklin H. Topp, article by Wilson G. Smillie. St. Louis: C. V. Mosby Co., 1952.
Malthus, Thomas Robert. *An Essay on Population.* New York: Dutton, 1958.
McVail, John C. *Half a Century of Small-Pox and Vaccination.* Edinburgh: Livingstone, 1919.

Stearn, op. cit.
Waterhouse, Benjamin. *A Prospect of Exterminating the Small-pox.* Cambridge: Cambridge Press, 1800 and 1802.
Winslow, op. cit.

PAMPHLETS:
"The 'Real Expedición Maritima de la Vacina' in New Spain and Guatemala," by Michael M. Smith. *Transactions of the American Philosophical Society,* Philadelphia. Vol. 64, part 1, 1974.
"Variola and Vaccinia: History and Description," by the New England Vaccine Co., 1894.
"Vaccination," by Joseph F. Edwards, Philadelphia, 1882.

CORRESPONDENCE:
Thomas Jefferson to John Redman Coxe, April 30, 1802; Thomas Jefferson to John Redman Coxe, Dec. 6, 1802.

Chapter IX
The Twentieth Century

BOOKS:
History of American Epidemiology, op. cit.
McVail, op. cit.

PAMPHLETS:
"Smallpox and its Prevention," New York Life Insurance Co., New York, 1915.

NEWSPAPERS:
New York Times: April 7, 9, 10, 13-19, 21-26, 1947.

DOCUMENTS:
"Epidemiologic Aspects of Smallpox in Yugoslavia in 1972," by S. Litvinjenko, B. Arsic, S. Borjanovic, Belgrade, World Health Organization, WHO/SE/73.57.

Memo to Director, Center for Disease Control, Atlanta, from Assistant to the Director, State and Community Services Division on Smallpox in Yugoslavia, Sept. 22, 1972, CDC EPI-SEP-72-91-2.
"Report on Yugoslavian Smallpox Conference," by Donald P. Francis, CDC, Dec. 1, 1972.

Chapter X
The First God to Fall

INTERVIEWS:
Benjamin Blood, telephone, 8/3/78; Larry Brilliant, Chelsea, Mich., 7/22/78; William Foege, Atlanta, 8/21/78; Stan Foster, Atlanta, 8/21/78; D. A. Henderson, Baltimore, 5/8/78; Ralph Henderson, Geneva, 5/30/78, and Atlanta, 6/11/76; George Stroh, New Delhi, 6/2/78.

Chapter XI
The Reluctant Warriors

INTERVIEWS:
Blood; D. A. Henderson; Victor Ladnyi, Geneva, 5/30-31/78; Benjamin Rubin, telephone, 9/2/78; Nedd Willard, Geneva, 5/29/78; Pierre Ziegler, New Delhi, 6/2/78.

Chapter XII
E^2 and the Needle

INTERVIEWS:
Malcolm Bierly, telephone, 9/2/78; Foege; Foster; Nicole Grasset, Geneva, 5/30/78; D. A. Henderson; Rubin.

MISCELLANEOUS:
"A Brief History of the Development of the Bifurcated Needle for Smallpox Vaccination," by B. A. Rubin, manuscript.
"The History of Smallpox Eradication," by D. A. Henderson, manuscript.
Patent, United States Patent Office, No. 3,194,237.
"Selective Epidemiologic Control in Smallpox Eradication," by William H. Foege, J. Donald Millar and J. Michael Lane, *American Journal of Epidemiology,* Vol. 94. No. 4, 1971.
"Smallpox Eradication in West and Central Africa," by William H. Foege, J. D. Millar and D. A. Henderson, *Bulletin of the World Health Organization,* Vol. 52, 1975.

CORRESPONDENCE:
D. A. Henderson to Malcolm Bierly, 8 March, 1967; J. H. Brown to D. A. Henderson, 18 June, 1970; D. A. Henderson to J. H. Brown, 24 June, 1970; J. H. Brown to D. A. Henderson 22 Feb. 1971; D. A. Henderson to H. Tint, 11 May 1972; H. Tint to D. A. Henderson, 23 May 1972.

Chapters XIII & XIV
[India]

INTERVIEWS:
R.S. Bajpai, Lucknow, 6/7/78; R. N. Basu, New Delhi, 6/5/

78; Blood; Brilliant; Foege; Foster; Mary Guinan, Atlanta, June 11, 1976; Grasset; Ladnyi; Lane; A. Monnier, Geneva, 5/30/78; Orenstein; Alan Schnur, Geneva, 5/26/78 and 5/29/78; M. I. D. Sharma, Delhi, 6/6/78; Bagamber Singh, Lucknow, 6/7/78; R. P. Singhal, Lucknow, 6/7/78; John Wickett, Geneva, 5/26/78; Willard.

MISCELLANEOUS:
"Clinical Observations on Smallpox: a Study of 1,233 Patients Admitted to the Infectious Diseases Hospital, Calcutta, 1973," by D. N. Guha Mazumder, S. De, A. C. Mitra, and M. K. Mukherjee, *Bulletin of the World Health Organization.*
"Cure-Deities of Bengal," by Asutosh Bhattacharyya, *Folklore,* Vol. 3, 1962.
"Current Status of Smallpox in the World," by Donald A. Henderson. *Journal of Communicable Diseases,* Vol. 7, no. 3, 15 Aug. 1975.
"Eradication of Smallpox in India," Report of the International Assessment Commission, 6-10 April, 1977, WHO, SEA/Smallpox/78, 4 Aug. 77.
"Genesis, Strategy and Progress of the Global Smallpox Eradication Programme," by Donald A. Henderson. *Journ. Comm. Dis.,* Vol. 6. no. 3, New Delhi, Sept. 1974.
"History of Achievement of Smallpox 'Target Zero' in India," by M. I. D. Sharma and Nicole C. Grasset. *Journ. Comm. Dis.,* Aug. 1975.
"The Last Known Outbreak of Smallpox in India," by Z. Jezek, M. Das, A. Das, M. L. Aggarwal, Z. S. Arya. *Indian Journal of Public Health,* Calcutta, Vol. 22, no. 1, Jan./March, 1978.
"Mission Possible: Death for a Killer Disease," by Lawrence and Girija Brilliant, *Quest/1978,* May/June 1978.
"National Smallpox Eradication Programme in India—Progress, Problems and Prospects," by M. I. D. Sharma,

William H. Foege and Nicole C. Grasset. *Journ. Comm. Dis.,* Sept. 1974.

"Operation Smallpox Zero," by Z. Jezek, R. N. Basu. *Ind. Journ. P. H.,* 1978.

"Operational Guide for Smallpox Eradication in India," Government of India/World Health Organization, Jan.-March 1975.

"Outbreaks of Smallpox in Delhi," by. R. R. Arora, D. Phukan, T. Verghese and A. Das. *Journ. Comm. Dis.,* Sept. 1974.

"Popular Beliefs about Smallpox and Other Common Infectious Diseases in South India," by R. J. Mather and T. J. John, *Tropical and Geographica Medicine,* Vol. 25, 1973.

"Report of the WHO International Commission on Assessment of Smallpox Eradication in Indonesia," WHO, April 25, 1974.

"Smallpox Eradication in India." New Delhi: Government of India/World Health Organization, 1977.

"Smallpox Surveillance Status in India," by. R. N. Basu. *Journ. Comm. Dis.,* Sept. 1974.

"Songs of the Goddess Shitala: Religio-Cultural and Linguistic Features," by Indira Y. Junghare. *Man in India,* Vol. 55, no. 4, 1975.

"Special Programme of Smallpox Searches Conducted Among the Floating Population of Calcutta," by Beverly Spring. *Journ. Comm. Dis.,* Aug. 1975.

"Studies of Thermostability of Lyophilized Smallpox Vaccine," by C. L. Sehgal. *Journ. Comm. Dis.,* Sept. 1974.

"A Study of the Goddess Sitala," research paper by Linda Shapiro, Univ. of Massachusetts, Dec. 15, 1977.

"War Against Smallpox—Development of Stable Vaccine," by P. K. Topa. *Journ. Comm. Dis.,* Sept. 1974.

NEWSPAPER:

"Study Finds a Certain Stability in Life of Calcutta Street

People," by Lawrence K. Altman, *New York Times,* Oct. 9, 1975.

CHAPTER XV
BANGLADESH

INTERVIEWS:
Andy Agle, Geneva, 5/30/78; Basu; Brilliant; Foster; Grasset; D. A. Henderson; Schnur; Wickett.

NEWSPAPERS:
"Asia Not Free of Smallpox Yet—9 Cases Found in Bangladesh," by Lawrence K. Altman, *New York Times,* Nov. 22, 1975.
"Global War on Smallpox Expected to Be Won in '76," by Lawrence K. Altman, *New York Times,* Sept. 29, 1975.
"WHO Takes Back Statement That Asia Is Free of Smallpox," *Wall Street Journal,* Nov. 23, 1975.

MISCELLANEOUS:
"Bangladesh Smallpox Eradication Programme," World Health Organization, WHO/SE/76.88, 1976.
"The Current Status of the Bangladesh Smallpox Eradication Programme," Government of Bangladesh, May 7, 1975.
East Pakistan Report, CDC, I.E.4-b, 1967, manuscript.
"Smallpox Eradication in Bangladesh." Government of Bangladesh/World Health Organization. Dacca: 1977.
"Surveillance in the Eradication of Variola Major from Bangladesh," by Stanley O. Foster, Kenneth Hughes, Joarder Abdul Kashem and Daniel Tarantola, presented 8th International Scientific Meeting of the International Epidemiological Association, Puerto Rico, Sept. 17-23, 1977.

Chapter XVI
The Long Last Stand

INTERVIEWS:
Isao Arita, Geneva, 5/21/78; Abdullahi Deria, Mogadishu, 6/26/78; Foster; Zdeno Jezek, Mogadishu, 6/26/78, and Merka, 6/28/78; Ali Maow Maalin, Merka, 6/28/78; Schnur; Wickett.

CORRESPONDENCE:
K. L. Weithaler to J. Copland and I. Arita, Feb. 28, 1975; Chief, Administration and Finance to K. L. Weithaler, Oct. 7, 1975; R. F. Lavack to K. L. Weithaler, Jan. 27, 1975.

MISCELLANEOUS:
"Ethiopia: Development of the Programme since 1971," WHO news release. SME/INF. 2.
"Last Known Smallpox Outbreak in Somalia," summary of investigation and measures taken. Z. Jezek, World Health Organization. SOM-N3 (undated).
"A Report from Gemu-Gofa Province, Ethiopia," by Ato Girma Talahun, Ato Kassa Mondaw, Daniel Kraushaar, Scott Holmberg. WHO/SE/72.48.
"Report on the Helicopter Accident," May 3, 1976. WHO.
"Smallpox Surveillance in Ethiopia," *Monthly Report,* Jan. 1975, Government of Ethiopia.
"Smallpox Surveillance in Ethiopia," *Monthly Report,* May 1975.

Epilogue

INTERVIEWS:
Arita; Lawrence and Girija Brilliant; Foege; Wickett; Willard.

NEWSPAPERS:
Associated Press, Sept. 2, 7, 1978.
Philadelphia *Inquirer:* Aug. 27, 1978.
Reuters, Aug. 30, 31, 1978.
Times of London: April 6-11, 16-18, June 19, 23, 26, 27, July 1, 5-6, 10, 17-19, 26, 1973.

MAGAZINE:
"Biological Warfare Fears May Impede Last Goal of Smallpox Eradicators," by Nicholas Wade, *Science,* Vol. 201, July 28, 1978.

INDEX

Authors Guild Backinprint.com Editions are fiction and nonfiction works that were originally brought to the reading public by established United States publishers but have fallen out of print. The economics of traditional publishing methods force tens of thousands of works out of print each year, eventually claiming many, if not most, award-winning and one-time best-selling titles. With improvements in print-on-demand technology, authors and their estates, in cooperation with the Authors Guild, are making some of these works available again to readers in quality paperback editions. Authors Guild Backinprint.com Editions may be found at nearly all online bookstores and are also available from traditional booksellers. For further information or to purchase any Backinprint.com title please visit www.backinprint.com.

Except as noted on their copyright pages, Authors Guild Backinprint.com Editions are presented in their original form. Some authors have chosen to revise or update their works with new information. The Authors Guild is not the editor or publisher of these works and is not responsible for any of the content of these editions.

The Authors Guild is the nation's largest society of published book authors. Since 1912 it has been the leading writers' advocate for fair compensation, effective copyright protection, and free expression. Further information is available at www.authorsguild.org.

Please direct inquiries about the Authors Guild and Backinprint.com Editions to the Authors Guild offices in New York City, or e-mail staff@backinprint.com.

9 780595 168675